PRINCIPLES
OF QUALITY CONCRETE

PRINCIPLES
OF QUALITY CONCRETE

PORTLAND CEMENT
ASSOCIATION

JOHN WILEY & SONS, INC.

NEW YORK   LONDON   SYDNEY   TORONTO

Copyright © 1975 by Portland Cement Association

All rights reserved. Published simultaneously in Canada.

No part of this book may be reproduced by any means,
nor transmitted, nor translated into a machine language
without the written permission of the publisher.

Library of Congress Cataloging in Publication Data

Portland Cement Association.
 Principles of quality concrete.

 "First in a series of five textbooks that comprise the
National concrete technology curriculum."
 Includes index.
 1. Concrete I. Title
TA439.P82   1975   624'.1834   74-17358
ISBN 0-471-67434-6

Printed in the United States of America

10 9 8 7 6 5 4 3 2 1

ACKNOWLEDGMENTS

This series is a national program to educate persons for employment in the concrete industries sponsored by the Portland Cement Association, the National Ready Mixed Concrete Association, and the American Concrete Institute in cooperation with the Office of Education, U.S. Department of Health, Education and Welfare.

PROJECT STAFF

George R. White,
   Project Director
Raymond J. Padvoiskis,
   Senior Editor

David Anderson,
   Associate Editor
William Perenchio,
   Technical Advisor

TECHNICAL REVIEWERS

The staff acknowledges the many excellent contributions by outstanding people in the concrete field and the advisory group that made up the administrative team. Of special significance were contributions by Howard Wiechman, John Seeger, John Metcalf, Dr. J. R. D. Brown, and others who helped launch this undertaking. Special mention goes to the instructors and students who tested the trial materials at the pilot schools and whose comments and suggestions have been incorporated in these volumes.

DISCLAIMER STATEMENT

This work was developed under a grant from the U.S. Office of Education, Department of Health, Education and Welfare. However, the opinions and other content do not necessarily reflect the position of the agency, and no official endorsement should be inferred.

This publication is based on the facts, tests, and authorities stated herein. It is intended for the use of professional personnel who are competent to evaluate the significance and limitations of the reported findings and who will accept responsibility for the application of the material that it contains. The Portland Cement Association and publication cosponsors disclaim any and all responsibility for the application of the stated principles or for the accuracy of any of the sources other than work performed or information developed by the Association.

Caution: Avoid prolonged contact between unhardened (wet) cement or concrete mixtures and skin surfaces. To prevent such contact, it is advisable to wear protective clothing. Skin areas that have been exposed to wet cement or concrete, either directly or through saturated clothing, should be thoroughly washed with water.

PREFACE

Concrete structures are a great deal more than sand, gravel, cement, and water stirred up and left to harden into usefully shaped lumps. Considerable care and knowledge are required to produce quality concrete before it is even placed.

Section One deals with the people who have an important share of responsibility for quality concrete—the concrete technologists. A technologist's general responsibilities and job opportunities are described. The section also reviews the history of the use of cement and some of the concrete products.

Section Two discusses the materials that make up concrete and describes their characteristics and requirements. Aggregates, for instance, are examined in terms of durability, resistance to freeze/thaw cycles, fire resistance, types, qualities, and classifications, processing and handling, and methods of determining soundness and absorption.

Section Three examines the general properties of quality concrete and gives a more precise definition of the chemical composition, workability, durability, and strength of concrete and the factors that affect these characteristics.

The final section deals with proportioning and mixing operations, which are critically important in making economical concrete that is satisfactorily workable and that, when hardened, has all of the specified qualities needed for the job. The various factors to be taken into account in trial mixes are discussed, and the proper methods of mixing are fully explained.

Principles of Quality Concrete is the first in a series of five textbooks that comprise the National Concrete Technology Curriculum. The U.S. Office of Education funded the preparation of these instructional materials to assist in the construction industry's continuing need for the development of personnel with a fundamental knowledge of concrete technology. The other books in the series are *Basic Concrete Construction Practices, Special Concretes and Concrete Products, Administrative Practices in Concrete Construction,* and *Concrete Inspection Procedures.* The Series is supplemented by a *Laboratory and Exercise Manual on Concrete Construction* and *Instructor's Guides.*

Portland Cement Association

TABLE OF CONTENTS

PRINCIPLES
OF QUALITY CONCRETE

SECTION ONE
CONCRETE TECHNOLOGY

CHAPTER 1
INTRODUCTION AND JOB OPPORTUNITIES

The business climate of the 1970s reflects increased population growth, rise in the Gross National Product, soaring construction volume in most sectors, and a technological revolution in building techniques. Environmental concerns, consumer evolution, and changing work incentives all add to the shift of emphasis from pure technical knowledge to a humanization of our ultimate work and life goals.

GROWTH OF TECHNOLOGY

At one time in industrial history, it was possible for a person to enter into a trade or occupation only by training on that particular job, without the benefit of any formal studies. Today, this is not the case. While technological growth has been significant to industry for over a century, it has gained particular importance since the end of World War II. Technological change, in addition to establishing new products and industries, has affected the character of labor itself. Many existing occupations requiring limited education and technical "know-how" have either been eliminated or have declined in importance; other occupations requiring additional education and technical knowledge have been created.

3

The U.S. Department of Labor, in a series of Manpower Research Bulletins, describes technological change as any change in the method of producing or distributing goods or services resulting from the direct application of scientific or engineering principles. Included in their listing of technological developments are new production techniques and resources, new methods of handling materials, the use of new and substitute products, and more efficient information dissemination and managerial control.

Technological change has brought about change on social, political, and cultural levels—changes continually reshaping the economic structure of the United States.

THE ECONOMY

Our economy may be considered to be a technologically oriented one in that the introduction of any new technology affects all other factors comprising our national well-being. The complexity of this economy has resulted in an increasing emphasis and need for technical workers of all kinds.

THE TECHNICIAN

Technology is the practical combination of the principles of science and engineering. The *technician,* as the worker in the various fields of industrial technology is known, *is the possessor of practical performance skills based upon scientific and engineering principles.* The technician has gained occupational recognition in the past two decades as a vital part of the scientific and engineering team responsible for the growth of our industrial technology. Technicians, a relatively young group of workers, are able to make more and more important additions to our present technological knowledge.

Theory, in order to be useful to mankind, must be put to practice. The work of the technician focuses on the practical rather than the theoretical side. Although considered by some to be limited in scope, this utilitarian aspect is what will create and maintain the demand for qualified technicians in the future.

The practical assistance of a technician balances and complements the theoretical work of scientists and engineers. Many of the technician's jobs would otherwise be performed by a scientist or engineer. The technician must therefore be not only manually skilled but also be technically knowledgeable and adept.

Basically, there are five general areas that the technologically trained graduate may enter. (See Fig. 1-1.) They are (1) translation of concepts into usable forms (research and development); (2) coordination of production facilities; (3) production itself; (4) preparation of public acceptance for new ideas and techniques; and (5) development of final products.

NEED FOR TECHNOLOGICAL TRAINING

The first element for success—opportunity—is ready-made for a graduate with technological training. Technicians are needed throughout the cement, concrete, and related industries. A very early report by Wickenden and Spahr brought out the need for additional education to supplement the work of the engineering profession. The opening sentence of this report stated:

> "A need exists in our post-secondary scheme of education for a large number of technical schools, giving a more intensive and practical training than that now provided by engineering colleges."

A later National Science Foundation study estimated that the need for technicians of all kinds would exceed 1,250,000. The present government

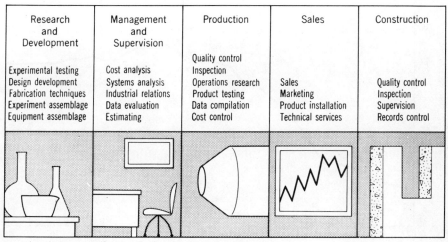

Research and Development	Management and Supervision	Production	Sales	Construction
		Quality control		
Experimental testing	Cost analysis	Inspection		
Design development	Systems analysis	Operations research	Sales	Quality control
Fabrication techniques	Industrial relations	Product testing	Marketing	Inspection
Experiment assemblage	Data evaluation	Data compilation	Product installation	Supervision
Equipment assemblage	Estimating	Cost control	Technical services	Records control

Figure 1–1   Work areas—concrete technologist.

estimates are that approximately 200,000 technicians of all kinds will be needed each succeeding year.

In viewing the various kinds of technicians required by the national economy as a whole, it was found that engineering technicians, the general category to which most concrete technicians belong, accounted for over half of the national requirements. A 78% increase in technician demand (73% for engineering technicians) is expected, the result of continued economic expansion, especially in those areas where growing complexity of new products and processes will stimulate the demand for highly trained personnel.

Illustrative of the continuous growth being experienced in industry are the figures for total cement usage in our economy (Fig. 1-2).

The wider use of technicians in expanding research and development activity historically follows the economic growth of a country. The largest source of supply for the needed technicians will be the graduates of curricula in industrial technology.

In considering the manpower needs of our national economy five years ago in comparison with what they will be five years from now, an obvi-

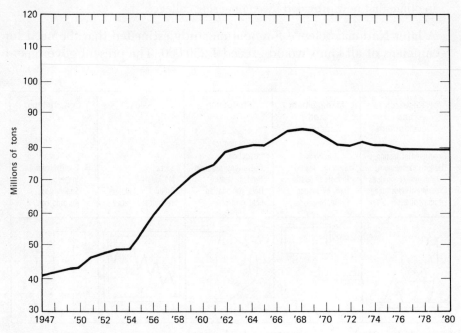

Figure 1-2   Cement-producing capacity in the United States (millions of tons—1947-1980).

ous growth pattern is indicated. Table 1-A has taken into account all major occupational groups. Compared to the size of the working force in 1965, projections show that in 1975, the professional and technical groups will comprise 140% of this number. This is the highest percentage rise of all the occupations. Of all occupational groups, only the farm occupations will require fewer workers. This is a good illustration of what a growing technology has done to the national labor force.

Dr. Harry N. Huntzicker, past president of the Portland Cement Association, speaking in regard to the cement and concrete construction industries, stated that the need for technician training is even more imperative in these industries than in many others.

"... the fact remains that the industry is still organized largely along handicraft lines. We are doing things too much as we did fifty years ago. We are still not ready to take our place in an industrialized construction industry that must perform near miracles in the next three decades if society's needs are to be met. Why is this so? Why are we not ready? ... There are no industry-wide programs of education and training of foremen, craftsmen, and sales personnel."

To educate is to create a problem-solving capacity. To train is to develop skills. Students entering the field of concrete technology will receive an education and training enabling them to take advantage of the vast opportunities that are already in existence and that are expected to enlarge in scope in the very near future.

Table 1–A
NATIONAL MANPOWER NEEDS BY 1975

Occupational Group	Percent Change in Numbers Employed (1965–1975)
Professional, technical	+40
Managers, officials, proprietors (except farm)	+25
Clerical workers	+33
Sales workers	+20
Craftsmen, foremen	+25
Operatives (semiskilled)	+15
Service workers (barbers, hospital attendants, waiters)	+40
Farm occupations	−25

Scientific and technological requirements for workers in the concrete industries have become much more demanding the past few decades. When concrete was a material with limited uses, it was not necessary to know exactly how it would react. Now, concrete is used to cantilever floors 100 stories above ground. Knowing the exact strength of the concrete is a matter, therefore, of utmost concern. There was a time when it was generally assumed that a smooth surface meant satisfactory concrete. Present requirements of concrete design, reinforcing, and prestressing are not satisfied merely by smooth surface appearance. Greater demands are placed on workers engaged in the concrete industries, and such demands can only be satisfied by greater education and training. With the growing trend toward certification, testing, and inspecting, this area of technological training is of paramount importance.

The sincere effort that an individual puts into any undertaking is governed by the amount or type of reward that is expected. In a career situation, the rewards are both the wages and the satisfaction obtained through doing a job well. The following statement is contained in the booklet *Technical Education May Be for You*, published by the Sun Life Assurance Company of Canada:

> ". . . the average laborer in the construction industry earned only ⅔ as much as a skilled tradesman in the same industry. In dollars and cents, the prospective craftsman can expect to earn 30% more in his working lifetime than his unprepared friend who quit school. According to current estimates, the technician can expect to earn about $100,000 more in his employment lifetime than the untrained or semi-trained."

Within 5 to 10 years after starting work, a concrete technologist earns more money than 70% of all working Americans.

At present, many unskilled and semiskilled occupations are becoming obsolete on the labor scene. Other occupations are in a process of redefinition.

The ability and thoroughness that a technician brings to his work has a significant effect on the profitability of his efforts in the cement and concrete industries. Concrete technicians must have the ability to learn and apply basic engineering and technical principles and methods, have the facility to know and use mathematics, and be able to communicate well. One of the primary requirements any technician must have is a continuing interest in current technology. A student's ease in comprehending various portions of the course of study may be indicative of his interests.

In some instances, a concrete technician's training will enable him to fulfill an employer's requirements for a production foreman or for a salesman. In the research and development area, many specific duties may be required.

In the areas of management and supervision, surveys conducted on the concrete technology curriculum show that prospective employers seeking administrative assistants would prefer the background that graduates of this program would have.

On an actual concrete construction project, a concreting foreman:

"Supervises and coordinates activities of work crews engaged in preparing and applying concrete for fabricating, covering, and reinforcing structures—buildings, bridges, highways, and dams; oversees workers who place forms for molding concrete, install reinforcing steel, and convey, place, finish, and cure concrete; reads specification sheets for information, such as sizes of aggregates and proportions of cement and water required, to insure that concrete is mixed according to specifications; inspects bracing and shoring of concrete forms to insure their stability before and during placing; indicates positions for placing chutes and runways, cranes, or paving machines to facilitate conveying concrete from mixer to forms; directs workers who spread, vibrate, screen, and float concrete to insure that concrete is compacted to desired consistency and surfaces are finished to specified uniformity and smoothness; and examines concrete after forms are stripped and gives instructions to workers to repair defects such as roughness and honeycombed appearance."

Apart from performing the supervisory duties of a concreting foreman, the technician is of great importance in the quality assurance and inspection of concrete construction. The overall job description for a construction inspector reads as follows:

"Inspects and oversees construction work to insure that procedures and materials comply with specifications. Measures distances to verify accuracy of dimensions of structural installations and layouts. Verifies levels, alignment, and elevation of installation, using surveyor's level and transit. Observes work in progress to insure that procedures followed and materials used conform to specifications."

One survey was directed entirely at cement producers. Although the specific job titles differ somewhat in the various answers, several distinct job classifications are used by cement manufacturers. The broad classifications and titles follow:

Sales: Sales Representative, Special Sales Representative, District Salesman, Technical Sales Representative, Sales Management, Sales Engineer.

Technical Services: Concrete Service Technician, Technical Serviceman, Technical Service Engineer, Field Service Engineer, Technical Assistant, Concrete Engineer, Technical Services Manager.

Laboratory Services: Laboratory Assistant, Laboratory Technician, Chemical Analyst, Chemist.

Production: Quality Control Technician, Concrete Technician, Quality Controlman, Production Supervisor.

Having considered both the manufacturer of cement and the producer of concrete products, there remain the users of concrete. Information was gathered from companies engaged in the paving industry about career possibilities for graduates of a concrete technology curriculum. The result:

". . . a concrete paving contractor could use a man with this training to take complete charge of all the material connected with concrete paving and also control our mix and mix design."

CHAPTER 2
CEMENT AND CONCRETE HISTORY

WHAT IS PORTLAND CEMENT?

Portland cement is a carefully proportioned combination of lime, silica, iron, and alumina. These materials are ground, thoroughly mixed, burned at a temperature of 2700°F, and the resultant clinker is ground and powdered. A new substance—concrete—is formed when this cement powder is mixed with water, aggregate, and sand. It has the advantage of going through a formable, plastic state and then curing to a hard, rocklike state.

The calcium of the formula is found in limestone, shells, chalk, or marl. The silica is frequently found in shale, clay, slate, blast furnace slag, silica sand, and iron ore. The other ingredients are found almost everywhere. Fortunately these raw materials are abundant and many areas have one or more of each type readily available. "Cement rock," rock containing the necessary elements of cement in a ratio approximating the formula of portland cement, is less abundant. These rocks can be crushed, burned, and ground directly into a cement powder that may or may not have the desired qualities.

Portland cement accounts for 98% of all cement sold in the United

States, but there are other types of cement that either have been made or still are being made.

Pozzolan cement, very similar to Roman cement, is still produced.

DISCOVERIES: PREMODERN ERA

The history of portland cement is the history of man's search for building materials superior to those provided by nature. Early man lived in an extremely hostile world. Only by using his powers of reason and unique opposing thumb was man able to develop weapons and tools. One of his first concerns was the construction of shelters. Anthropologists and archaeologists believe that early man lived either in natural caves or in crude huts built of twigs, mud, and rocks. Such primitive structures could not have been very satisfactory. They were uncomfortable, unstable, and unsafe.

While rock is a fine basic building material, its value depends on the mortar used to bind it into a permanent structure. Very early cultures used mud. The Babylonians and Assyrians employed natural deposits of bituminous material; the Egyptians grooved building blocks, fitting them together.

An archaeological excavation at the Uruk digs in Iraq in 1968 unearthed a templelike structure. The building, tentatively dated from 4000 B.C., was partially constructed of concrete—its earliest known application.

Later Egyptian cultures, in addition to fitted building blocks, utilized a cementlike, lime-and-water mortar to bind stones used in larger structures. The Great Pyramid is constructed of gypsum stone bound with crude gypsum mortar (Fig. 2-1).

The Romans were the first to use cement mortar. Examples of early Roman architecture in which cement mortar was used include the pre-Christian baths, the Coliseum, and the Basilica of Constantine. A mixture of slaked lime, volcanic ash (Pozzolana), and pottery shards resulted in a Roman cement that was both durable and hydraulic; unfortunately the Roman process disappeared with the empire that had devised it. A permanent, hydraulic, construction-binding substance did not appear again until the eighteenth century (Fig. 2-2).

THE SEARCH FOR BETTER CEMENTITIOUS MATERIAL: MODERN ERA

When searching for a building material with which to rebuild the Eddystone Lighthouse in 1756, British engineer John Smeaton discovered

Figure 2–1 The Great Pyramid is constructed of gypsum stone bound together with a crude gypsum mortar—one of the early uses of a prepared plastic cement.

Figure 2–2 The Coliseum—one of the structures in which the Romans used cement mortar.

hydraulic cement. That structure withstood the fury of the Atlantic Ocean for over 125 years.

In his work, *Hydraulic Mortars, Etc.* (1869), Dr. Wilhelm Michaelis stated, "A century has elapsed since the celebrated Smeaton completed the building of the Eddystone Lighthouse. Not only to sailors, but to the whole human race is this lighthouse a token of useful work, light in a dark night. In a scientific point of view it has illuminated the darkness of almost two thousand years.... The Eddystone Lighthouse is the foundation upon which our knowledge of hydraulic mortars has been erected, and it is the chief pillar of modern architecture." Whether or not Michaelis overstated the case, he was correct in calling Smeaton's contribution "the chief pillar of modern architecture." (See Fig. 2-3.)

Figure 2–3 The use of hydraulic mortar for constructing the Eddystone Lighthouse in 1759 marked the rebirth of a long-forgotten technique.

In 1796, Joseph Parker converted nodules (commonly referred to as "noodles") of argillaceous limestone found in the London area into cement. This type of cement, later called "Roman" cement because of its similarity to the mortars made by the early Romans, was used to build the Thames River Tunnel. In 1822 James Frost was granted a patent on a substance he called "British" cement. A mixture of limestone or marl containing silicious earth or silica was ground and calcined in a kiln until "all carbonic acid be expelled and until it be found on trial of a small portion of such calcined materials that it will not cool, slake or fall when wetted with water." Frost was quite specific as to the makeup of the raw materials for his products. In 1825, he built the first cement factory at Swanscombe, England (Fig. 2-4).

Joseph Aspdin, a bricklayer and mason from Leeds, England, was granted a patent in 1824 for a cement he called "Portland" for its similarity to the stone found on the Isle of Portland. Aspdin's patent called for a mixture of limestone and clay ground together and heated until all

Figure 2–4 By 1908, 83 years after Frost built the first cement factory, his factory in Swanscombe, England had expanded until it looked like this.

carbonic acid had escaped. Aspdin's patent reads, in part: "My method of making a cement or artificial stone for stuccoing buildings, waterworks, cisterns, or any other purpose to which it may be applicable (and which I call Portland Cement) is as follows: I take a specific quantity of limestone, such as that generally used for making or repairing roads, and I take it from the roads after it is reduced to a puddle or powder; but if I cannot procure a sufficient quantity of the above from the roads I obtain the limestone itself, and I cause the puddle or powder, or the limestone as the case may be, to be calcined. I then take specific quantity of argillaceous earth or clay, and mix them with water to a state approaching impalpability, either by manual labor or machinery. After this proceeding I put the above mixture into a slip pan for evaporation, either by the heat of the sun or by submitting it to the action of fire or steam conveyed in flues or pipes under or near the pan, till the water is entirely evaporated. Then I break the said mixture into suitable lumps, and calcine them in a furnace similar to a limekiln till the carbonic acid is entirely expelled. The mixture so calcined is to be ground, beat, or rolled to a fine powder, and is then in a fit state for making cement or artificial stone. This powder is to be mixed with a sufficient quantity of water to bring it into the consistency of mortar, and thus applied to the purposes wanted."

Aspdin is recognized as the father of modern portland cement by many, but not all, authorities. Joseph Parker's patent described a material much closer to the portland cement of today. Aspdin used the name "portland" but the product was, in the words of one of his contemporaries, ". . . . no more like cement that is made today than chalk is like cheese." It differed from portland cement in several important ways. The raw materials were not heated nearly as hot as for modern cement, and no definite proportions of raw materials were mentioned.

Aspdin was a most colorful character who would carry a copper tray into his plant and mysteriously scatter various powders over the raw materials being ground for firing, much to the interest of other masons and builders. He built a cement plant at Wakefield, England, in 1828, which produced, along with Frost's plant, cement used in the construction of the Thames River Tunnel.

John Bazley White, proprietor of one of the new cement works, manufactured the cement for the London sewer system built between 1859 and 1867. It was in 1837, however, that White commissioned an architect to design and build the first all-concrete house. I. C. Johnson, a general manager of the White works, experimented with Aspdin's cement. He discovered the need for careful firing of the raw materials to a point of semivitrification but not beyond and the need for an exact combination of raw ingredients. These discoveries were a major turning point in the development of the industry. He also designed a chamber kiln capable of heating raw materials for the next firing.

In 1855, construction started on a favorite project of William Aspdin, son of the famed Joseph. Young Aspdin designed a house to demonstrate the beauty possible with the use of portland cement. Unfortunately, the plan proved too costly, and after the house ("Portland Hall") was only one-third completed, he was forced to abandon the project. It is interesting to note, however, that in 1925, the completed portion was still standing and the concrete work did not show any great amount of weathering. As early as 1848, *The Builder*, a trade magazine, reported the test results of the various types of these early cements. Testing began early in this industry—and it is still going on today.

While these developments were being worked on in England, the rest of Europe was not sitting idly by. Both the Germans and the Belgians were busy innovating and improving, as were the French. By the late 1880s, the Belgians were so advanced in cement production that they supplied nearly all the cement imported by the United States. In 1877, the German Portland Cement Manufacturers Association was formed with 29 member companies producing 2.4 million barrels of cement annually. By 1903, member companies had grown to 94 and production to 22 million barrels.

PORTLAND CEMENT IN THE UNITED STATES

The development and growth of the cement industry in the United States closely parallels the growth of the country itself. As the population grew and began moving westward, it created a need for better and safer transportation. Farm goods and raw materials had to be shipped to the cities and manufacturing centers of the East. Because water transportation was the easiest and most economical method available, a series of man-made canals were planned to connect existing natural waterways. Planning for the Erie Canal began in 1817. In 1818, the U.S. cement industry was born when Canvass White discovered a natural cement-producing rock in Madison County, New York. Cement from White's plant was used to construct the Erie Canal.

Shortly after White's discovery, other deposits of "cement rock" were located in Rosendale, New York, the Lehigh Valley of Pennsylvania, and

Louisville, Kentucky. The construction of other canals—the Chesapeake and Ohio, Lehigh and Pennsylvania State, and Great Canal (near Louisville)—provided ready markets for the new cement producers.

Shipments of portland cement started entering the United States in 1868. Used as ballast in ships, cement was transported at almost no cost. By 1895, over 550,000 tons of European cement were being imported to the United States annually. U.S. production at the time was only 90,000 tons.

The first portland cement plant in the United States was built by David O. Saylor in 1870. Located at Coplay, Pennsylvania, the plant began operations the following year. The poor quality of Saylor's early product was corrected by John W. Eckert, the plant's chemist. Eckert was able to analyze the problem as incorrect proportioning of raw materials. His adjustments resulted in a consistently high-quality cement good enough to be displayed at the Philadelphia Centennial Exhibition of 1876 (Fig. 2-5).

In the early 1870s in South Bend, Indiana, Thomas Millen and his sons were producing concrete pipe with cement purchased from Knight, Bevan & Sturge, an English cement plant. One day a stranger passing by the Millen works noticed the KB&S cement barrels in the yard and, having worked for the English plant, introduced himself to Thomas Millen. In the course of their conversation he asked Millen why cement was purchased from England when the South Bend area abounded with the

Figure 2-5 David O. Saylor built this first cement plant to operate in the United States in 1870.

necessary raw materials. Millen immediately located a book describing the cement-producing process and experimented with local marl and blue clay. The first batches were burned in a sewer pipe (perhaps the first rotary kiln ever used), and the clinker was ground in a coffee grinder. The experiments proved successful and within six months the Millens had their first kiln in operation; they were soon able to add three more. They found that they could not manufacture cement fast enough to satisfy the demand and later combined with other manufacturers to build a larger production plant.

By 1874, the industry had grown enough to warrant the beginnings of a cement market. Robert W. Lesley founded Lesley and Trinkle, a firm of cement brokers. He later entered the manufacturing field and developed several time- and labor-saving devices for the production of cement.

Farther west, the Alamo Cement Company was founded in San Antonio, Texas, in 1880. The first plant was next to the county poorhouse causing skeptics to suggest that the founders would not have a long trip when their foolish scheme failed. The company sold a great deal of building stone, and, although their cement sales were poor, they managed to survive and avoid that trip next door. To create increased demand for cement they started building concrete sidewalks. When a sidewalk slab was placed and troweled, they covered it with planks to protect the finish—and to secretly check the quality of their product before unveiling it for public approval. The company was successful, and, in 1908, the demand for its products was so great it expanded and was renamed the San Antonio Portland Cement Company (Fig. 2-6).

THE AMERICAN PORTLAND CEMENT INDUSTRY GROWS

By the middle of the nineteenth century, the full impact of the Industrial Revolution was felt in the United States. Not an integral part of the American scene prior to the Civil War, manufacturing grew and flourished during that period as factories struggled to supply armies on both sides of the conflict. The nation, primarily agricultural before the war, continued to industrialize afterward. Farm workers in great numbers began seeking employment in factories.

The total energy of the United States seemed to channel itself into building. Railroads were laid as fast as men could cut and trim the ties. Buildings were higher, bridges were longer, and transportation was faster. Growth was both inward (to the cities) and outward (to the West). Sophistication in engineering and architecture required new

Figure 2–6   The Alamo Cement Company was founded in San Antonio, Texas in 1880.

strength and durability in building material. The result was the development of concrete.

PRODUCTION PRACTICES

There were three early methods of producing portland cement.

The wet method was used for chalk-and-clay mixtures. During the crushing and grinding of the raw materials, water was added to insure good mixing. The resulting sludge was allowed to settle, and the residual matter was first air-dried, then furnace-dried, and finally fired in a kiln mixed with layers of fuel.

The semiwet process, used mostly in France, called for a complete grinding of raw materials before the addition of water to aid in the mixing of these ingredients. The mixture then passed through rollers to extract as much water as possible, was air dried, and finally fired in a kiln.

The dry method used no water at all in the grinding. Raw materials were ground and mixed together. In some processes, water was added to this pulverized mixture so that briquettes could be formed before firing.

Until the 1870s, U.S. cements were all natural cements; the only portland cement available was imported from Europe. The importation of cement reached its peak in 1895 when almost 565,000 tons of cement entered the United States. But by 1897, more than 50% of the cement used in this country was U.S.-produced. While the natural cements of the Rosendale and Lehigh districts enjoyed a fine reputation, the introduction of American portland met with a great deal of consumer resistance. As a result, many of these first American portlands were marketed under foreign labels. Between 1897 and 1899, U.S. production of portland cement doubled. The Spanish-American War offered the portland cement industry a market, but U.S. portland still had two thriving rivals—the American-produced natural cements and the foreign portlands. Nevertheless, by 1900, the production of portland cement in this country surpassed that of natural cements.

In the early 1900s, the U.S. portland-cement producers were paying their laborers from one-quarter to one-half more in wages than were being paid to European workers. It was imperative that cement be produced more economically; this could best be done by decreasing the labor force. One major breakthrough in the solution of this problem was the development of more economic kilns. The first cement had been fired in dome, or bottle, kilns. It was necessary to charge the kiln with each firing. That is, for each batch of raw materials being burned the kiln had to be filled with fuel. At the end of the firing period it had to be cooled before the contents could be emptied. The results were wasted fuel, lost heat, and nonuniformly burned materials. To solve some of these problems the vertical kiln was developed. Saylor's plant at Coplay used vertical kilns. The vertical or shaft kiln was a definite improvement, but many problems still remained.

Rotary kilns were first used in England in 1887. Frederick Ransome patented an improved rotary kiln. This kiln was tilted slightly out of horizontal plane and slowly rotated. The material moved from one end to the other down the incline of the kiln and could thus be kept in constant motion. This kiln had a much greater capacity and burned more evenly than previous models. (Fig. 2-7).

As described in *Cement* by Bertram Blount:

"This kiln consists of a cylindical furnace, set at a slight inclination, carried on rollers, and rotated by worm-gearing. At the upper end, powdered raw material is fed in by the hopper and travels down the

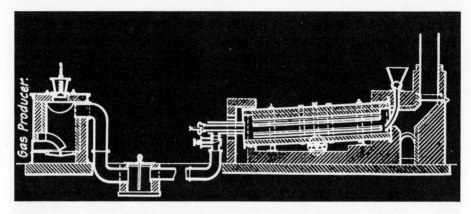

Figure 2-7   The original rotary kiln: Ransome's patent of 1885.

furnace, meeting burning producer-gas entering by a pipe. The burnt material falls into a pit, and any raw material which may be blown back up the furnace is caught in the settling-pit, whence it can be returned to the hopper."

In 1890, Jose F. DeNavarro visited Ransome in England, obtained the American rights to this kiln, and so U.S. manufacturers were able to capitalize on this major improvement.

Thomas A. Edison expressed an interest in the portland cement industry as early as 1880; however, taking advantage of the low-yielding iron ore found in New Jersey, Edison opened an iron-producing plant at New Village, New Jersey, in the 1890s. When the rich Minnesota Mesabi Range was discovered about the same time, Edison converted his operation to cement production. The challenge to improve the industry was irresistible and his inventive genius was responsible for a number of innovations. These included improved grinders for the pulverization of both raw materials and clinkers, electrically powered equipment for heavy work, and rotary kilns over 150 ft long.

One basic problem faced by the early cement industry was quality control. In 1904, the American Society for Testing Materials (now the American Society for Testing and Materials) advanced some specifications for portland cement that were adopted by many of the cement producers. These specifications have since been revised periodically. With production capacity increasing, quality control became more and more important to the individual producer.

The growing production capacity of American cement plants (the 1902 output was over 25 times that of 1894) caused overproduction to become a problem. A greater market had to be found. The portland cement industry was most fortunate to have men who combined the talents of salesmanship with a firm belief in the product they were selling. It was up to these men to sell concrete to the public, and they did. Wherever concrete could do a better job than other materials, a cement-company salesman appeared and sold.

In 1891, the first concrete street was laid in Bellefontaine, Ohio. It served that town for 60 years before resurfacing was necessary. The amazing success of the concreted portion of this and other early streets led to the construction of concrete streets and sidewalks in cities all over the country. Lincoln Highway, now known as U.S. 30, was constructed shortly after World War I. Although this coast-to-coast highway was not completely paved with concrete, cement companies contributed enough cement to build concrete "seedling miles" on the highway and a 3-mile concrete stretch near Dyer, Indiana. This 3-mile section set the precedent for concrete road construction all over the United States during the great highway construction programs after 1920.

During the early 1900s, the federal government entered the dam-building business. Flood control became more and more important as people settled near rivers and as deforestation increased runoff into major rivers. Hydroelectric power was, of course, an important by-product of these dams. Dams, levees, and dikes all require concrete for construction. The United States also completed the Panama Canal, started in 1880 by a French company. The canal, which used great quantities of concrete for locks and other integral parts, opened in 1914.

As the country continued to expand and become more sophisticated, there was an increased need for concrete. Grain elevators, stadiums, sewage plants, sewer pipe, concrete-masonry construction, sidewalks, building foundations, smoke stacks, water towers, water-supply systems, office buildings, factories, schools, and many other major necessities depended on concrete.

By 1920 all the elements necessary for a major expansion of the cement and concrete business were present. This expansion was reflected in the sudden appearance of several allied industries.

One of the earliest uses of concrete and mortar was the construction of water and sewer pipelines. An adequate supply of clean water was essential to the life of any town or city. In some areas, an abundance of natural wells and streams provided ample water sources. However, in more densely populated areas, where intracity water resources were insufficient, clean water had to be transported from surrounding regions.

Centuries before the birth of Christ, the Romans constructed a hydraulic-cement waterline—basically tamped concrete pipe—that ran from the Eiffel Mountains to what is now Cologne, Germany. It remained in service until 1928. Since pressure pipe capable of carrying pumped water was not developed until the late nineteenth century, all water systems prior to that time depended on gravity and wheel/lock devices. These early systems remain extraordinary engineering feats even today.

Another problem faced by the early city dwellers was that of sewage disposal, critical to the health of any city. In the ancient Sumerian city of Nippur, ruins of a sewer arch constructed about 3740 B.C. have been found, as have the remains of a sewer found in Tell Asmar near Baghdad, Iraq, dated from around 2600 B.C. (Fig. 2-8). About 1700 years before Christ, the Minoans on Crete built stone sewers and drains. In 200 B.C. the Romans built the *Cloaca Maxima*, a large drain constructed of tamped concrete pipe that was used until the early 1900s. These ancient peoples partially solved their problems.

During the turbulence of the Middle Ages, knowledge of the production of hydraulic cement was lost. It is interesting to speculate if and how world history might have been altered had the various plagues that ravaged Europe in this period been avoided by adequate sewer systems.

In 1850, the city of Paris constructed a sewer system of rock that was heavily coated on the inside with cement (Fig. 2-9). The first-known concrete-pipe sewer in the United States was laid in Nashua, New York, in 1842. Concrete pipe was used for sanitary sewers in the 1840s, and concrete drain tile and storm sewers were introduced about the same time. Concrete irrigation pipe was first used in California shortly after the 1849 gold rush.

Concrete pipe won wide acceptance in sewer construction after the Civil War. In 1868, sewers were installed in St. Louis, Missouri. The pipe was 15 in. in diameter and was made by the hand-tamp method. Over 95 years later, these sewers were still in excellent condition and

Figure 2–8 In the Sumerian city of Nippur ruins of a sewer arch, constructed about 3740 B.C., are still visible.

showed every indication of serving the community for a long time to come.

Although early concrete pipe was made of nonreinforced concrete, as larger and stronger pipe was required, reinforcing became a necessity. In 1870 Jean Monier, a French gardener and one of the first to reinforce concrete with steel, received a patent on a reinforced, concrete pipe. An Italian engineer named Mazza developed asbestos-cement pipe in 1913. A combination of cement, asbestos fibers, and silica, the pipe was ideally suited for water distribution systems, sewage and waste-water collection and disposal systems, irrigation and drainage for agriculture, cable conduit, vents for heating and air-conditioning units, as well as other uses. It was first manufactured in England in 1925 and four years later was introduced to the United States.

A new type of concrete pressure pipe was developed in the 1920s—

Figure 2–9 In 1850, the city of Paris constructed a sewer system of rock that was heavily coated on the inside with cement.

thin steel cylinders totally coated with cement. By 1930 prestressed concrete pipe reached the market. The original design has subsequently been improved upon, and prestressed pipe capable of withstanding great pressures has been developed. Over the years there has always been a certain amount of cast-in-place concrete pipe. This method is used in especially large installations and can be seen on jobs demanding unusual shapes and pipe of a size too large to transport easily.

THE CONCRETE-MASONRY INDUSTRY

Concrete masonry has existed as long as portland cement. Buildings of concrete were built before concrete walls were cast. The first machine designed to make concrete masonry was marketed in 1895. One contractor who did not have a suitable molding machine had a carpenter build forms and then sent employees to the city dump to collect tomato cans. Two cans equaled one core. The resultant block was reported to be most satisfactory.

During the early 1900s, concrete masonry was often used for decorative purposes as well as building. Some of the early concrete block was cast to resemble rough-hewn granite. Buildings from this era are still seen in every city and town. Decorative cornices and building blocks were also widely used.

During the years between 1920 and 1940, the concrete-masonry industry grew significantly and received wide public acceptance. The use of concrete masonry at first was subjected to a great deal of prejudice by builders and building-code authorities. The doubts about it were reflected in the inflated fire insurance rates that concrete-masonry-constructed buildings had to pay in the 1920s. As a result, the Cement Products Association, American Concrete Institute, and Portland Cement Association sponsored and directed a series of laboratory tests on the material. The results of these tests proved conclusively the fire resisting and retarding qualities of concrete masonry. Later tests establishing the great load-bearing qualities of the product further enhanced concrete masonry's reputation as a superior building material.

In 1920, Francis J. Straub of Pennsylvania was granted a patent for the fabrication of concrete block from cinder aggregate. The inventor cited the advantages of insulation capability, sound absorption, texture, nailability, and lightness of the resulting block. Concrete masonry made with other lightweight aggregates (Haydite, Waylite, Celocrete, Superock, and volcanic cinder and pumice aggregates) was introduced shortly thereafter.

In 1930, the Concrete Products Association became the Concrete Masonry Association and, two years later, published the first definitive work on the merits of concrete masonry, *Facts About Concrete Masonry*. The First Concrete Industries Exposition was held in 1937, in conjunction with the annual convention of the National Concrete Masonry Association. For the first time, producers of the machinery used to make concrete masonry brought their wares to an industry meeting.

The parallel growth of the concrete-masonry industry and general improvements in concrete capability and production can be dated from the mid-1930s. Labor-saving automation devices were introduced regularly—machines capable of mixing, forming, transporting, curing, loading, sorting, and checking concrete masonry. By the mid-1950s, fewer plants were producing more masonry. The shift to lightweight aggregates for concrete masonry production extended to the point where today lightweights are the rule rather than the exception.

The use of concrete block for decorative purposes grew rapidly after 1960. At first used almost exclusively in warmer climates, the lacy blocks have become a prime architectural tool for producing aesthetically functional building units.

The concrete-masonry producers, through the NCMA, have sought to coordinate and standardize their products. Their efforts have been successful, and the concrete masonry industry enjoys the reputation of producing masonry that is easily usable.

During World War II, the block industry worked at top capacity to produce the building materials so desperately needed by government and essential support industries. After the war, production increased further as hundreds of new concrete masonry plants appeared across the nation. Many of these new plants were owned and managed by returning servicemen who found a good market for their products in the home construction field.

During the post-World War II years, great attention was given to improving the product. Block-producing equipment design advanced markedly, and more attention was given to the curing of the cast block at temperatures of 180°F or lower and using steam at low pressure. The interest in high-pressure curing by steam is rising. Block is placed in an autoclave and subjected to pressures of 140-150 psi at 350°-365°F. Two or three loads of block can be cured in a single day using this high-pressure steam method.

The concrete-masonry industry grew as the equipment to produce and handle the block was developed. Many concrete masonry plants are highly automated and virtually all block is produced under pressure and vibration in high-speed automatic molding machines. Almost complete automation is possible in the concrete-masonry factory of the future. By eliminating as much costly labor as possible, it is economically feasible for the producer to charge less for his product—and this price advantage makes concrete masonry more and more attractive to the consumer.

It is predicted that the use of concrete masonry will increase in the years to come. What was once a rather haphazard industry has grown to be one of the most respected. Quality control, improved design and strength factors, increased beauty, and comparative low cost have made concrete masonry a favorite with builders and architects. As with other allied industries within the concrete and cement field, concrete masonry producers today are just beginning to realize their full potential.

THE READY-MIXED CONCRETE INDUSTRY

In 1909 the residents of Sheridan, Wyoming, could have witnessed the birth of the ready-mixed concrete industry had they noticed the horse-drawn wagon carrying the first portable cement mixer. Ironically, the wagon's destination was a job site where an automobile garage floor was being placed. The turning of the cart's wheels powered the transit mixer.

Prior to World War I, concrete was prepared in stationary plant mixers and hauled to construction sites in dump trucks. The practice of delivering premixed concrete seems to have originated in Baltimore, Maryland; however, two Oklahoma companies were delivering 1000 cu yd of concrete annually in this manner as early as 1900.

Many delivery truck designs were tried. Some trucks were equipped with agitators to keep the mix from settling during transit; others were equipped with mixing units on each truck. Stephen Stepanian applied for a patent for a ready-mix truck (Fig. 2-10) in 1916, but the patent was not granted.

In 1921, a Milwaukee firm attached a mixer to a truck and mixed concrete en route to the job site; the load was discharged by gravity. All of the early truck mixers were made with the axis of the drum horizontal. However, in 1937 a mixer appeared that had the discharge end of the drum elevated. These high-discharge or inclined-axis mixers permitted chuting the concrete into forms higher and further from the mixer unit. This innovation was quickly accepted, and today very few horizontal-axis mixers are produced. In 1937, mixer capacities were generally between 2 and 5 cu yd. Because of rising labor costs and other factors, the size of truck mixers has steadily increased. Today, the most-common sizes are 7 to 9 cu yd with a few 15- and 16-cu yd models being manufactured each year. During the first years of ready-mix operation, only one type of concrete was available, but as batching (the gathering to-

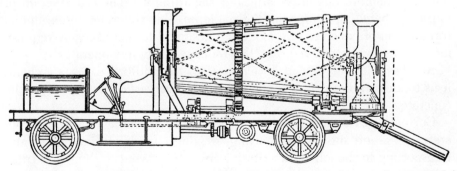

Figure 2-10. In 1916 Stephen Stepanian applied for a patent for a ready mix truck, but it was not granted. The drawing represents the design that he envisioned.

gether and measuring of ingredients necessary to produce a batch of concrete) improved, greater diversity in the concrete mixture became possible.

Since 1930, the ready-mixed concrete industry has been striving for greater efficiency in operation and increased quality control of their product. The National Ready Mixed Concrete Association was established to help formulate standards for the industry, disseminate new ideas and developments, and to act as the industry-representative body.

The most interesting aspect of the ready-mix industry is its fantastic growth. In 1925, there were 25 ready-mix operators in the United States. By 1929 this number had grown to 100, and today there are more than 6000. From this growth pattern it is obvious that the ready-mix plant had something to offer. The consumer wanted his concrete delivered to the job in a ready-to-place condition. This growth has been accompanied by improvements in the equipment and methods used by the ready-mix operator. Early central-mixing plants compounded their batches of concrete by volume—so-many-measures of cement to so-many-measures of aggregate to so-many-measures of water. This system provided no way to determine the amount of moisture in the sand, and when this quantity was not taken into account, two problems arose. First, the water content of the sand could vary to a great extent and cause mixes to be either too wet or too stiff. Second, the volume of the sand varied significantly with moisture content. As engineers and architects found more specialized applications for concrete, their specifications became more exacting. More accurate methods for determining mix proportions were needed. Today, ingredients are measured by weight. (The moisture content of the sand is determined by a moisture meter, and the amount of water introduced into the concrete mixture is adjusted accordingly.) The introduction of improved batching equipment has made possible the exact and scientific formulation of concrete mix. Concrete designed to meet specific requirements can now be ordered directly from the ready-mix operator.

Another relatively recent development is the addition of certain chemicals to a concrete mixture to alter and improve various properties of the concrete.

Currently, there are two basic methods of providing ready-mix concrete to the job site: *central-mixing*—mixing completely in a plant and transporting to the job in an agitating truck to prevent settling, and *truck mixing*—mixing completely in a truck mixer. A combination of these two methods, in which concrete is partially mixed in a plant and then fully mixed in a truck mixer, is referred to as *shrink-mixing*.

Other equipment used by the ready-mix operator has been improved. Bins and silos are used for the proper storage of aggregates and cement; conveyor equipment moves dry materials to the mixing site. There are improvements such as the pneumatic movement of cement, heated bins, improved weighing devices, accurate dispensing of admixtures, auto-mated batching, electronic moisture meters, and improved cranes. The specifications for cement, aggregates, and admixtures have become more selective over the years to meet new requirements.

The ready-mix concrete industry uses over half the production of all U.S. cement plants. Contractors as well as homeowners have discovered the economies possible with job-delivered plastic concrete. Whether it is the foundation of a high-rise building or a one-lane driveway that is being placed, the ready-mix truck is usually on the job.

THE PRECAST-CONCRETE INDUSTRY

Since its discovery concrete has been used to precast specialty items. Paving blocks, window sills, steps, highway guardrail posts, survey mark-ers, patio blocks, trash burners, bird baths, auto parking curbs, concrete burial vaults, and other items have all been precast for many years. In the past two decades, however, this branch of the concrete industry has grown into one of the most important. A group of remarkably creative men has developed design concepts utilizing precast concrete members that are revolutionizing the building industry. Another group, equally creative, has taken the basic idea of precast concrete and turned it to fine art.

In both the practical and aesthetic branches of the precast field, suc-cess depends to a large extent on the talents and abilities of the workman. As new designs and new forms are attempted, it is only through the vision and capabilities of the skilled craftsman that producers can meet the specifications of the architect or artist. In many areas of this expand-ing field there are no set rules or definite methods; the trial-and-error method must often be used.

Although almost unknown in the United States until about 1950, in Europe, British engineers were experimenting with precast residential structures as early as 1905. Following World War II, precast construction became a permanent part of the European building trades. The devasta-tion wrought by the war demanded a plan of rebuilding that would at once be extensive and rapid. The answer was precast system construction.

Precast building systems fall into four major categories: post and beam, panel and slab, box, and component.

Post and beam systems, including frame systems, consist of prefabricated columns and beams with wall panels. Wall panels and floors may either be on-site or factory-precast. The overall result of such systems is great flexibility for interior design.

Panel and slab systems use large floor and wall components, assembled on the site, to form cross-wall, exterior-bearing wall, or core-bearing structures. The panels themselves are either precast on the site or at the factory. Panel and slab construction was developed in Europe and constitutes the largest segment of systems construction in that part of the world. The British Bison system and the Camus system, developed by the French, are the two most popular European methods of panel-and-slab construction.

Box systems include all systems that utilize three-dimensional components. A box system of construction is just that—construction using entire boxes (rooms) rather than panels and frames. Habitat, constructed for Canada's Expo '67, is a notable example of this type of system (Fig. 2-11).

Figure 2-11 Habitat, constructed for Canada's Expo '67, is a notable example of the "box system" type of construction.

Component systems involve the use of factory-produced components—complete wall assemblies, kitchen and bathroom core units, plumbing and heating units, and the like. In component-systems building, the double "T" system, developed by the Colorado Prestressers Association, is one that is getting wide acceptance and use.

Of the four types of precast systems, the panel-and-slab system was developed in Europe and is used exclusively there; the other three systems are products of North American engineering, and their use is limited to the United States and Canada.

Precast structural members are used in nearly every field of construction. Pilings and decks for railroad and highway bridges, beams for retaining walls, railway crossties, beams, girders, and concrete piles (some of them as high as a 10-story building) are all in common use today. The use of precast concrete in a structure can often offer decorative benefits at a relatively low price. Architects are using this medium with increasing frequency. Exposed aggregate panels, high- and low-relief finishes, textures, contrasting-colored concrete, and the use of decorative aggregates all contribute greatly to the variety that can be obtained in precast members. Each general method offers countless variations.

The Polish Institute of the Warsaw Enterprise for Industrialized Construction has developed a process of gluing gravel to paper, then making a sandwich of the gravel-coated paper, mortar, and load-bearing concrete. The process is reportedly 14% cheaper than previous methods and takes only one-third as long.

Students of the fine arts are using precast concrete for bas-relief sculpture. One popular method is the carving of a styrofoam mold. This plastic substance enables the sculptor to create fine edges, and the styrofoam imparts an interesting texture to the finished panel. Other precast work is used to create interesting architectural detail in buildings. Talented men are just discovering and perfecting some of the many things that can be done with this plastic, durable, and strong material.

REINFORCED CONCRETE

Another area of cement and concrete is the field of reinforced concrete. Concrete is extremely strong in compression but has little tensile strength —about 10% of its compressive strength. That is, a concrete member when placed on-end can support a great deal of weight, but the same beam when used horizontally to span a gap cannot support nearly as much. This fact limited the early uses of concrete to those jobs where only compressive strength was necessary.

W. B. Wilkinson was perhaps the first person to reinforce concrete when, in 1853, he used wire rope to reinforce a floor slab. In 1865, Francois Coignet used longitudinal and transverse rods to reinforce floors. The first patent for concrete reinforcing was granted to Jean Monier, the previously mentioned French gardener and a man of extreme versatility. In 1867, he was granted a patent by the French government for the embedding of wire in concrete to add strength to the finished product. (He was making flower pots.) Later, Monier extended his original patent to include the manufacture of reinforced concrete pipe, floors, beams, and bridges. Although he was not, strictly speaking, the first to make use of reinforcing materials, Monier did much to popularize reinforced concrete. He eventually built thousands of reinforced concrete gas and water tanks all over France.

In the 1890s, Francois Hennebique developed a method of using two sets of reinforcing rods—one in the lower layer of the concrete slab and a set in the upper layer. For a distance near the center of the span the upper rod drops down to almost touch and parallel the bottom rods. With a few modifications, this basic system is still in use today. Hennebique was the first builder to utilize integral construction—the casting of slabs, beams, and columns as a single unit.

In the 1870s, Thaddeus Hyatt did some testing of the expansion and contraction coefficients of steel and concrete. He found them to be almost the same. He was thus able to prove that a bond developed between concrete and steel during the curing of a piece of reinforced concrete; his tests proved that this bond would not break down. About the same time, William Ward commissioned architect Robert Mook to design the first reinforced concrete house. Ward deduced, through his own experiments, that iron would be able to give the greatest amount of support to horizontal beams if it were placed near the bottom of the beams. He used reinforced concrete for all walls, beams, floors, cornices, and towers. Doubting neighbors immediately labeled the structure "Ward's Folly," but when it did not collapse they changed the name to "Ward's Castle."

Ernest L. Ransome (son of Frederick Ransome, developer of the first rotary kilns) did a great deal of building using reinforced concrete. He introduced a system of construction using columns and floors to carry the weight of a building. In this way, walls could be much thinner, and valuable floor space could be saved.

Throughout the first 20 years of this century, a great deal of testing and checking of reinforced concrete was done. Scientists and engineers subjected reinforced concrete to every test their minds could devise. The testing, of course, continues. The American Concrete Institute publishes a great many manuals and booklets of interest to concrete workers. One of them, the ACI *Standard Building Code for Reinforced Concrete* (ACI 318), gives minimum requirements for design and construction of reinforced concrete. ACI is dedicated to providing member organizations with up-to-date information on the properties and uses of concrete.

THE PRESTRESSED-CONCRETE INDUSTRY

Prestressing is one method that engineers have devised to compensate for concrete's low tensile strength. Unlike reinforcing, the steel embedded in prestressed concrete is either put in under tension (pretensioning) or it is inserted in ducts after the concrete has hardened and then stretched (posttensioning). The tension in the steel causes a balancing compression in nearby concrete. When loads begin to be imposed on the beam, this compression in the concrete must be relieved before tension appears.

Peter H. Jackson may perhaps have received the first patent for prestressing concrete when, in 1888, he was granted a U.S. patent for the "construction of artificial-stone or concrete pavements." The methods he described in the patent application are startlingly similar to posttensioning techniques in use today. During the 30 years that followed, many men came up with various methods of pre- and posttensioning, but it was not until 1919 that Karl Bernhard Wettstein devised a way to mass-produce prestressed concrete in plants. Prestressed concrete was used in Europe for some time before American designers started creating structures utilizing prestressed members. In 1949 the first prestressed concrete structure was erected in the United States—the Walnut Lane Bridge in Philadelphia (Fig. 2-12).

In 1939, C. C. Sutherland of John A. Roebling and Sons Division of the Colorado Fuel and Iron Corporation designed and supervised the construction of a warehouse floor using prestressed concrete members.

Eugene Freyssinet, a French engineer, was one of the men who contributed most to the field of prestressing concrete. In a letter dated 1951, Freyssinet wrote:

"I would like to state—as I used to say in the past when much less attention was given to my ideas than today—that the essential characteristics of prestressed concrete to which its remarkable properties are

Figure 2-12   In 1949 the first prestressed concrete structure was erected in the
United States—the Walnut Lane Bridge in Philadelphia.

due do not depend on prestressing of steel reinforcements, neither on
previous compression of concrete in the tension zones, because both
conditions can exist at least to a certain extent, without attaining the
essential properties of prestressed concrete. These properties are uniquely
due to an *exactly determined intensity of compression in the concrete,
to the condition of inequalty whch lets re-enter all possble strains* within
the elastic margin. These are essential characteristics which I have
expressed in 1928 after my patient efforts of emphasizing an idea conceived already in 1903, an idea that I have investigated experimentally
with a tensioned tie in prestressed concrete."

By providing tensile strength to an already strong building material,
prestressing has made concrete all the more attractive to contractors. The
Prestressed Concrete Institute, established in 1954 for promotion of architectural use of precast and prestressed concrete, represents the majority
of U.S. and Canadian producers.

CHAPTER 3
CONCRETE AND CONCRETE PRODUCTS

INTRODUCTION

In making a serious study of the concrete industry, it is worthwhile to consider some of the past uses of the product as well as some of its more recent developments. Such consideration serves to create an awareness of where the industry has been and, more importantly, where it is going. As new uses for concrete are developed, new businesses and jobs follow. Automation is eliminating the more routine jobs in the field, but advanced designs are creating an increased demand for men who have a knowledge of concrete and can apply this knowledge to fit the industry's needs.

Because of the variety of concrete uses, a strict categorical study of the subject is very difficult. This chapter has been divided into nine major headings to facilitate overall comprehension of concrete use.

1. Paving
2. Architectural uses and housing
3. Agriculture
4. Water control
5. Water restraint
6. Irrigation
7. Drainage
8. Recreation
9. Miscellaneous uses

While there is a certain amount of overlap between some categories, each heading is treated as completely as possible.

PAVING

Walks, streets, roads and airports utilize some 20% of the output of the nation's cement factories. Whether for a narrow path or an eight-lane expressway, concrete has proved to be durable, economical, and safe.

Sidewalks may be a cast-in place ribbon of concrete or precast blocks placed as stepping stones (Figs. 3-1 and 3-2).

Figure 3–1   Cast-in-place sidewalk construction.

Figure 3-2   Precast sidewalk construction. Notice the vacuum-lifting device that the men are using to lift the precast blocks into place.

Parking facilities and drives reflect America's transformation into a nation on wheels. Decentralization of business has forced many people to depend on the private automobile as their primary means of daily travel. The family car goes to the office, bank, store, and church—the list is almost endless. And all of this means that parking space must be made available; most parking areas are paved with concrete. The once-lowly parking lot has actually become the "front door" to many public and private buildings. Architects now consider parking lots as a function in total site development (Fig. 3-3).

Concrete offers the architect the advantages of attractive appearance, economy, long service, and long-term value. By using one or more colors, concrete can also function as a traffic regulator.

Industrial driveways and parking areas designed to support heavily loaded trucks are an important part of any commercial facility. Industrial planners who are interested in long life and low annual cost of these areas use concrete as part of the total concept of the factory site.

Roads and streets, and their quantity and quality, are good indicators of a nation's degree of industrialization and of the living standards of its people. The U.S. system of expressways and primary and secondary roads is unequaled by any other nation. U.S. citizens are an itinerant people, changing jobs often and spending vacations touring, driving hundreds of miles on a weekend to visit friends or to see a football game. Concrete

Figure 3-3 This multilevel parking garage illustrates the pleasingly modern design that such a utilitarian structure can assume.

roads have served the country's needs for many years. Among their characteristics are:

1. Low maintenance cost.
2. Long life.
3. Clean, attractive appearance.
4. Proven performance over many years.
5. High surface traction.
6. Good light-reflectance values.

A good concrete road must have a good base; even the best concrete placed on a poorly prepared roadbed cannot last. Soil-cement is one method used to stabilize a road sub-base when the existing soil does not provide the required firmness or when the traffic on the pavement is projected to be especially heavy.

Road patching is essential to the maintenance of concrete roads. There are two reasons for resurfacing a road: to smooth slabs that have become roughened so that they no longer meet the requirements of rideability and to strengthen slabs that are not structurally adequate for modern traffic loads. Resurfacing is a process that demands careful attention to detail to insure proper bonding of new concrete with the old. Old surfaces must be roughened and swept very clean of any dust or extraneous material before a good bond can be attempted.

Badly broken pavement is removed, and the subgrade is corrected. Forms are then placed, the area is dampened, and the new slab is placed.

The term "slabjacking" is used to describe an operation known for many years as "mudjacking." "Mud" is no longer used to jack pavements. Finely ground limestone or sand and portland cement mixtures have replaced previously used mixtures. The principle of slabjacking is simple. Grout is pumped under pressure through a hole cut into the pavement creating an upward pressure on the bottom of the slab. This upward pressure raises the slab restoring it to its original position.

But it is not only on the surface of the road that concrete is used. When shoulders are built of expansive soils, gaps can occur between the edge of the paving and the shoulder. These gaps allow water to enter the subbase of the pavement to the extent that even the most porous subbase material cannot drain it away quickly enough. As a result, the frequent passage of heavy axle loads during rainy weather causes excessive water surging along the pavement edges. This causes erosion of the shoulder and, in extreme cases, of the sub-base. Soil-cement is one of the ways to stabilize the shoulder, although concrete shoulders have been used in a few states.

Most concrete streets also have concrete curbs and gutters. These are either integral or cast separately. Integral curbs are cast as part of the slab. They often have very low profiles—low enough to allow driveways to enter a street without breaking the curb. Separate curbs are generally 18 to 24 in. high and 5 to 7 in. wide at the top. They are usually built after the road pavement is placed. A lip curb is a low and flat integral curb often used on country roads to prevent erosion of the shoulder.

Bridges and grade separations can also be considered part of paving; they are essential to meeting the requirements of modern, high-speed roads. Highways are built on all types of terrain, and must meet various urban and rural requirements. As a result, bridges and grade separation structures must be included in any highway program. Normal, open-road flow of traffic will not be interrupted at the intersection of two highways if the roadways are separated by means of grade separations. Continuous and full capacity of both highways can be assured by adequate interchange roads and properly designed entryways to allow turning vehicles to join traffic without interference. These structures must not only satisfy the requirements of their particular sites but must also fit into the overall engineering and be usable at another location without much modification. In this respect concrete is a particularly advantageous building material. Because of its plasticity and adaptability, changes in structural type or architectural form can be made easily.

Expressways passing through congested residential or industrial areas where right-of-way is restricted are frequently depressed, or below-grade. Local traffic is routed on overpass bridges and on service roads parallel to the main highway. Access to the expressway is usually infrequent and accomplished by means of interchanges.

Backslopes under bridges and grade separations are often protected with concrete block, soil-cement, or concrete to prevent soil erosion. Bridge railings and handrails for pedestrian stairs can also be made of concrete.

Concrete traffic controls, long-lasting traffic markers and median barriers provide roadway safety. In a recent comparative study conducted in California, concrete median barriers inflicted less vehicle damage and received no damage to themselves. This test proved that concrete medians were safer for the traveling public and required no upkeep.

Tunnel construction relies heavily on concrete as a building material; the majority of highway tunnels are constructed entirely or partly of concrete.

Transportation, other than strictly automotive, depends heavily on concrete. Railroads utilize concrete bridges and tunnels and, more recently, prestressed concrete crossties. Concrete crossties are not particularly new themselves but the prestressing of these members to increase strength is a recent innovation. In 1960, the first major tests of prestressed crossties in the United States were conducted. European countries have done more experimenting and prestressed crossties have been proven successful (Fig. 3-4).

Railroads have also solved some of the problems repeatedly encountered with shifting subballast on their roadbeds. A mixture of water, sand, and cement is applied to the roadbed, and, when set, this grout stabilizes the ballast. In some cases the grouting mixture is applied to the roadbed under pressure to force the mixture deeper into the subballast.

Where railroads cross highways there has always been the dilemma of assuring sufficient space for the wheel flanges of the locomotive and cars while maintaining a relatively smooth ride for an automobile. The spaces between the tracks are sometimes filled with wooden planks but these

Figure 3-4   Concrete and steel railroad ties.

warp and wear causing a noisy and bumpy ride for vehicles. Precast concrete slabs have been used successfully to alleviate this problem.

Airport runways made of concrete serve many large airports. Grooving the surface of a concrete runway reduces the deadly hydroplaning that sometimes occurs on wet runways. This condition has been the direct cause of landing accidents. Spilled fuels and jet blasts do not damage concrete surfaces as they might other materials. Concrete is the most durable of all possible paving materials for airport runways. The increased visibility offered by its light color is an additional safety feature. Small airports, searching for a permanent, rigid, and inexpensive paving surface, use soil-cement.

ARCHITECTURAL USES

Before discussing the ways in which concrete is used in the construction of buildings, large and small, some understanding as to the basics of concrete component design is necessary. With few exceptions, buildings conceived and executed in the past were two-dimensional. Post-and-beam design and construction seemed the easiest way to fill man's eternal need for shelter. This type of construction has been adequate and expedient, and in many cases still represents the best method of design. Concrete shell roofs have allowed architects some relief from the restraint of planar limitations. The domes that dot the architecture of the past are indicative. Such variances from conventional design proved prohibitively expensive and were used only where cost was relatively unimportant. In 1923, Carl Zeiss, famed German manufacturer of optical equipment, designed and built the first concrete shell roof. This signaled the opening of an era of new freedom in architectural design. In the years since the construction of that barrel shell roof, the size, types, and shapes of concrete shells have multiplied. Shell roofs are currently used for such divergent structures as churches, service stations, airplane hangars, auditoriums, industrial buildings, water reservoirs, and stores.

In addition to their spectacular beauty, concrete shells often prove to be the most economical means of roofing buildings. Simplified design procedures and improved forming techniques have made them highly competitive with other roof systems long thought to be lowest in cost. One reason for this surprising economy is the structural action of shell roofs.

Shells derive their strength, and, consequently, part of their economy, from a basic and easily comprehended principle of statics—form as an important factor in the development of strength.

Hold a sheet of paper along one end and lift it from a table. It hangs limply from the points of support because it has practically no strength when cantilevered. Roll the sheet into a half-circle and hold it along one edge. Now it will not only cantilever but it will also support small weights such as paper clips. An analogy can be made between the flat sheet of paper and a beam of shallow cross-section. When straight, both are weak and incapable of appreciable spans because they resist loads by means of bending stresses only. When formed into an arc, however, bending forces are practically negated. The remaining forces acting within shells can easily be handled by small amounts of concrete and reinforcement. Such curvilinear shapes make shell roofs the most efficient method known for enclosing space with concrete.

The sheet of paper mentioned above is a small version of a barrel shell. Barrel shells are of two types—short and long. Long barrels are those with chord widths that are small compared to the span between supporting ribs. Conversely, short barrels have large chord widths in proportion to the span between ribs.

Functional requirements and architectural considerations are the determining factors in making the choice between long and short barrel roofs. For spans under 100 ft that require approximately uniform clearance between floor and roof, long barrels are usually most appropriate. Great room structures where vaulted ceilings are practical—auditoriums, churches, gymnasiums, concrete halls, and theaters—are best constructed with short barrel shells.

Folded-plate shell roofs are noted for their amazing spanning and load-carrying capabilities. Their ability to cantilever has been capitalized on in aircraft hangars, schools, stores, and industrial buildiings. In two-story buildings, the second-story floor slab can often be suspended from a folded-plate roof. There are three basic types of folded-plate shells: V-shaped, Z-shaped, and a modified W-shape. Triangular plates can be used to obtain three-dimensional action (Fig. 3-5).

There is an infinite variety of double curved shells available to the architect. The two most prominent types are shells formed by the rotation of a curve about an axis, the most common being a spherical dome and the other a translation of a curve along another curve.

Of the many types of shells, the translational form is possibly the one best suited to the creation of a wide variety of free-form shapes with formwork. For example, when curvature along two perpendicular axes is in opposite directions, segments of this form can be combined to form the pleasing groined vault.

Figure 3-5 Folded-plate roofs give a great deal of strength.

Any combination of curvatures can be used, with the rise in one direction completely independent of the rise in the other. If the form is parabolic, the surface can be described by a series of straight lines.

When curvatures along the two axes are in the same direction, a dome rectangular in plan is achieved. Such a shape offers construction ease because the profile at all sections along the axis is identical.

Hyperbolic paraboloids of this type use a minimum amount of material. In most cases, even with supports at the four corners, only mesh reinforcement is needed. Again the thickness of the shell is dictated primarily by construction needs. Hyperbolic paraboloids create interesting roof lines.

A new method of casting a dome-shaped concrete shell roof was introduced in 1968. A deformable plastic membrane is laid on the ground and sealed at the edges to prevent the escape of air. Steel mesh is placed on it, and high-slump, plastic concrete is in turn placed on the mesh. The plastic membrane is then inflated to the desired height and the concrete is allowed to set. The cured concrete is cut to create windows and doors; the overall construction is amazingly fast and cheap. It is interesting to consider the many applications that this technique of inflated concrete construction could have.

Prestressed concrete is another recent development in concrete construction. Concrete has high-compressive strength but lacks tensile strength. Tensile strength is gained by the addition of stressed steel to the concrete. This can be accomplished either before or after the concrete is placed.

Pretensioning is the casting of a concrete beam with stressed steel embedded in it. Lengths of high-strength steel wire are stretched in forms, and concrete is placed around them. When the concrete has cured, the forms are removed and the tensioning is released from the wire. This tension is transferred to the concrete itself by the strong bond that develops between concrete and steel.

When a concrete unit is to be posttensioned, the piece is cast with a hole in it through which steel cables or tendons can be threaded. End anchorages are attached and the steel is stretched and anchored to the ends of the beam. The force may be applied directly to the anchoring plates or it may be transferred to other beams abutting the member being posttensioned. The hole is then usually filled with grout. This method is used most often in long-span beams and where members are cast in place (Fig. 3-6).

Prestressed concrete is used in beams, roof slabs, decks, bridges, domes, folded plate roofs, girders, floor slabs, joists, columns, pipe, and wherever great tensile strength is required. The first use of prestressed concrete was in the construction of bridges and large buildings.

The future of prestressed concrete is especially bright. Architects, engineers, and designers are already proposing complete "packages" for the construction of large commercial buildings. There will be increased use of prestressed concrete shells for hangars, exposition halls, factories, and other major buildings. Larger prestressed concrete pilings will be available; prestressed concrete railroad ties will find increased accept-

Figure 3–6   This double-T construction illustrates the use of pre-stressed concrete units.

ance; and prestressed concrete transmission poles will replace the more usual wooden and steel utility poles. Dams, monorails, highway paving slabs, and concrete bridges with longer spans will all use prestressed concrete.

Tilt-up construction is one building method that does not require a large number of workmen at the building site. While not a new concept, tilt-up construction has been more widely used in the past few years. Whole wall or floor units or large segments of these units are precast on the ground at the job site. When the concrete is cured the panels are tipped into place with a crane and secured (Fig. 3-7).

This type of construction is used on the farm, in the construction of warehouses, storage structures, schools, homes, and wherever a relatively inexpensive method of building is desired.

Precast concrete panels are being used to a greater extent now than ever before. Sometimes these panels are prestressed, often they are extremely decorative. Precast panels are often used as "curtain walls."

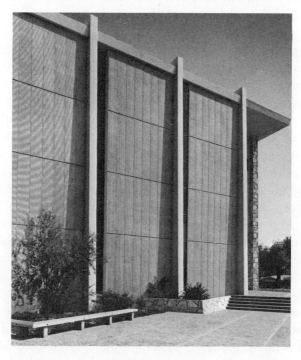

Figure 3–7 The textured surface on these tilt-up panels was created by the use of plastic form liners.

These walls supply insulation and low-cost protection from the elements as well as beauty but are not expected to carry the weight of the structure. Various textures possible with precast panels will be covered later in the chapter.

Precast construction applies the techniques of the production line to the precasting of concrete units, these contain one or more rooms or parts of rooms that when joined together form a building. The principles of precast construction have been used most often in raising apartment buildings. One method uses chemically prestressed concrete to form the interior and exterior walls and ceilings of an entire apartment. When the concrete has cured, the trim, cabinets, plumbing, electrical wiring, heating ducts or outlets, and other appurtenances are installed. The entire unit is delivered to the construction site (along with other such units) where a foundation has been prepared. A relatively small construction crew (six to eight men) can erect the entire building with the aid of a crane.

The U.S. government has shown a great deal of interest in this type of construction for urban renewal. The main problem of slum clearance has always been the relocation of the people living in the submarginal housing while the new apartments are being built. Usually years elapse between the time the old structures are razed and the new units readied for occupancy. Using precast construction, this time span is reduced to about one month. It would be possible to clear some land, build a precast building, move people into it, and clear the emptied tenements. This process could be repeated until a whole area would be cleared of slums; the residents of the area would be moved with a minimum of discomfort.

Concrete masonry constitutes another family of concrete products widely used in the construction and decoration of all types of buildings. The once-lowly block has, in the past years, received much attention and is now used for load-bearing and nonload-bearing walls, piers, partitions, fire walls, backup walls for brick, stone, and stucco-facing materials, fireproofing around steel columns, stairwells and enclosures, chimneys, garden walls, outdoor fireplaces, and many other uses. The utilitarian standard 8×8×16-in. block remains the backbone of the industry, but concrete masonry producers have developed other masonry that is beautiful as well as functional. Split block and slump block are used extensively for homes in many parts of the country. Grille walls of lacy "solar" block decorate many gardens; patterned and glazed face block provide interesting architectural textures to walls, both interior and exterior.

While much normal-weight aggregate block is produced each year, a trend toward block made with lightweight aggregates has developed. One of the most frequently used lightweight aggregates is expanded blast furnace slag. Cinders are also used to make cinderblock a popular lightweight concrete block. Extremely lightweight block is made using pumice, volcanic scoria or lava, perlite, or vermiculite as the aggregate. These various types of block have good acoustical and insulating qualities and are extremely fire resistant, however, they do not have the strength of the heavier construction block.

Some popular uses of concrete masonry include paving eating areas, construction of foundations and basements, lightweight partitioning, all types of interior and exterior walls, and garden walks and planters. Cement mortar used with concrete masonry is composed of portland cement, sand, water, and hydrated lime or lime putty. This mortar can also be used for repairing and tuckpointing old masonry walls.

A wide variety of designs and patterns may be produced in panels and in large tilt-up wall units by using textured materials as form liners. Vacuum-formed thermoplastics, fiberglass-reinforced plastics, plastics formed into shape by heat and pressure, treated wood, ribbed rubber mats, sheet metal, plaster of paris molds, and numerous other form liners can be used to achieve interesting results.

Once a concrete slab is formed, numerous treatments can be applied to change its texture. Bushhammering, a method of roughening the surface, is done with a pneumatic hammer that knocks off the surface cement paste exposing aggregate. The surface can also be sandblasted. This tends to leave a smoother surface than does bushhammering.

Architect Paul Rudolph created an especially pleasing texture for the Endo Pharmaceutical Center in Garden City, New York. Using vertically ribbed form liners to cast the walls and then hammering off the resultant ridges to a depth of one-half to two-thirds of an inch, a rough-hewn surface complementing the bold and massive forms of the building design was achieved (Fig. 3-8).

Architects and builders are using more and more exposed aggregate concrete on walls and paved areas. The variations of texture and color possible with this technique are limitless, and it is possible to carry the colors, forms, and materials of nature to modern building methods. There are several methods of creating an exposed aggregate finish. On precast panels, the aggregate-transfer method is often used. A layer of fine sand is placed in horizontal forms to a thickness of about one-third the diameter of the aggregate to be used. The aggregate is then pushed into the sand as densely as possible so that part of each piece is exposed. The

Figure 3–8 The Endo Pharmaceutical Center in Garden City, New York.

concrete and any reinforcing and/or insulating material is placed on top of this sand-aggregate layer, and the slab is allowed to harden. When the forms are removed, the sand is washed away leaving the aggregate embedded in the concrete.

Exposed aggregate finishes have been achieved by pasting the aggregate to form liners. The forms are filled with concrete, allowed to set, and the liners stripped away.

Marblecrete, an attractive finish that resembles exposed aggregate work, is prepared quite differently. Chips of marble are embedded into the final coat of a stucco or cement plaster wall, usually of white or colored cement. By using the marblecrete method it is possible to achieve the beauty and color of marble without the great cost (Fig. 3-9).

Colored concrete offers other architectural possibilities. Concretes produced with gray and white portland cements harden in the gray-white-buff range of colors. When pigments are mixed with cement the color range of concrete can be greatly extended. Integrally mixed pigments have been used for many years. White portland cement should be used for the mixing of light colors and pastels and for deeper hues where true color is desired.

A travertine texture can also be applied to slab-on-grade concrete. The concrete is placed and finished with a stiff broom. A finish coat of cement, sand, yellow pigment, and water is applied with a dash brush to create an uneven surface. Once this has started to dry, the ridges of the surface

Figure 3–9   An example of a marblecrete surface.

are flattened somewhat so that the final finish is a combination of smooth and rough.

A Russian architect has been glazing concrete surfaces with an oxy-acetylene torch. The extreme heat of the torch melts the minerals in the concrete to produce a ceramiclike glaze. Several different colors can be achieved by varying the oxygen ratio of the torch's fuel. While this technique is still in the experimental stages, it promises to offer a new possibility for the surface finishing of concrete. Precast bathroom and kitchen units could be glazed at the factory to give floors and walls the easy-to-clean, attractive finish of fine tile.

Terrazzo floors using marble or other colored aggregate chips set in concrete are usually made with white or colored portland cement. Placed on a concrete subfloor, the terrazzo is a relatively thin layer with 70 to 85% colored marble chips. This is allowed to set and is then ground and polished to a lustrous shine. Terrazzo will last as long as the building and has the decided advantage of never needing anything more than soap and water to maintain its beauty.

AGRICULTURE

The possibilities for the use of concrete around the city or suburban yard are limited compared to the number of uses the farmer or rancher has for concrete. Lack of fire protection in the country has led many farmers to use fire-resistant concrete or asbestos cement wherever possible. Concrete's durability, low upkeep, ease of construction, and adaptability make it additionally attractive.

Concrete masonry construction has long been used by the farmer. Other methods of concrete construction are being adapted to answer agricultural needs more and more.

Tilt-up construction methods may be utilized in barn or shed building. The labor necessary is minimal and flexibility in use is great. Once panels are cast and cured, the construction of a building proceeds rapidly.

Increased attention is being paid to "packaged" concrete buildings for farm use. One successful system uses precast columns for a roof support. With interior Y-columns, it is possible to add new sections of any desired width and length. The usual advantages of concrete are coupled with ease and speed of construction.

Much concrete work around the farm is cast in place. Concrete is helpful in a successful dairy operation and has been used for all types of animal shelters. It is fairly resistant to acids found in manure and is relatively easy to keep clean. In recent years farmers have developed more mechanized methods of handling manure. Below-grade concrete manure tanks are installed close to livestock facilities; manure from barns and paved areas is washed into sloping gutters that feed directly into this tank. Floors made of concrete slats allow manure to drop through into a gutter leading to the storage tank.

With concrete, barns can be kept cleaner and animals more comfortable. Less bedding is required, feed bills are reduced, and valuable manure is saved for fertilization. Concrete-paved barnyards can change a sea of mud into an area that is useful year around.

Long ago, farmers discovered the advantages of storing some of their corn and hay crops in silos; the fermented silage feeds their animals during the winter months. Several different methods utilizing concrete have been developed for improving silo construction. Some silos are cast in place using form sections that can be clamped together. Six to 8 ft of wall can be constructed in this manner each day. A more recent development is the use of a slip-form. The form is raised slowly and a rotating bucket continuously places concrete within the form until the silo is completed. Silos are also built of factory-made concrete staves held together with steel bands. Silos need not be vertical structures. A concrete-lined trough makes an adequate silo and has the advantage of low-cost installation. The farmer is often able to build it with little outside labor. The same type of horizontal silo can be constructed above grade using the tilt-up technique.

Concrete is especially valuable to the farmer in the construction of granaries and other grain storage structures. A concrete granary is vermin-proof and can save many bushels of grain each year. Various types of construction can be used: tilt-up, concrete masonry, cast-in-place, or fabricated panels.

Modern farming is highly mechanized. The available supply of farm labor has dwindled, and the farmer now depends more and more on machines. Dramatically demonstrated in the growing and harvesting of crops, it is just as true in the care and feeding of farm animals. Most new concrete silos are equipped with unloading augers; these are often connected to conveyor belts that deliver silage directly to concrete feed troughs. Dairy farmers find they can increase the size of their herds if they provide "loafing barns" for their cows and do their milking in milking parlors—both made of easy-to-clean concrete.

Concrete serves the farmer in the form of feed alleys in barns, feed troughs, mangers, pens, nonrotting fence posts, stock watering tanks, water troughs, and loading ramps. Hogs thrive in concrete farrowing houses where valuable sows and pigs are protected in a controlled environment. Concrete can also offer a pig the deluxe version of his favorite spot—the wallow. Most farmers will provide their hogs with a wallow or fog-cooled pen realizing that this is one way these nonperspiring animals have of keeping their body temperature at normal levels during extremely hot weather.

Concrete can assist the poultry breeder in the construction of vermin-proof houses and feed storage areas. Multistoried, caged laying houses are common to the poultryman who has an eye for continued profits.

WATER CONTROL

One of the major uses of concrete is in the restraint, transportation, conservation, and treatment of water. As an area of land is developed, water problems often occur. There may be water where it is not wanted, not enough where it is needed, or sometimes nature seems to attempt to drown everything. Concrete helps remedy these situations.

The health and economic well-being of every community and of the nation as a whole depend on adequate and satisfactory water supply for municipal, industrial, agricultural, and other uses. Water use by municipalities, industry, and agriculture increased from 40 billion gallons daily in 1900 to 260 billion gallons daily in 1955. The figure has continued to increase annually and is expected to double before 1980.

Presently, the greatest use of fresh water is for irrigation. While irrigation in the past has largely been confined to the western states, more and more acres in the other parts of the country are receiving supplemental irrigation to increase crop yields.

Pipe

The same qualities of durability, strength, economy and formability that make concrete so attractive to the architect, homeowner, highway builder, and farmer make it attractive to the hydraulic engineer. Many water-handling problems are solved by concrete pipe. There are three major use areas of concrete pipe: (1) water supply; (2) sanitary sewer, storm sewer, and culverts; and (3) irrigation. Two major types of pipe satisfy the requirements for all these uses: pressure pipe and nonpressure pipe.

Concrete pipe in a central water supply system is pressure pipe—pipe constructed to withstand specified internal pressures. As internal water pressure increases, the compression built into the pipe decreases so that the pipe operates at near zero stress. Most urban water systems deliver water under pressure other than the gravitational type provided by elevated water tanks. All pressure pipe has a tongue end and a groove end. Joints are finished with rubber that seals the tongue end of one pipe into the groove of the next one.

There are four main types of pressure pipe:

1. Low-pressure pipe is made with reinforcing of steel bars of mesh. Generally, it can withstand pressures of about 50 lb per sq in. (psi). It is often used for irrigation systems.
2. Reinforced cylinder pipe are steel cylinders, reinforced with steel mesh either inside or outside and coated with concrete. Larger-pressure pipe (6 to 7 ft in diameter) are usually made in this manner.
3. Prestressed cylinder pipe are concrete cores with steel cylinders on

the outside or inside, wrapped with a high-tensile wire. This makes a very strong pipe capable of withstanding 300 psi of pressure. It is most frequently used in water lines.

4. Pretensioned cylinder pipe is similar to prestressed cylinder pipe, but steel bars rather than high-tensile wire are used as the prestressing agents. Not as much stressing is achieved, so strength is lower; however, pretensioned cylinder pipe has the advantage of being lighter in weight. This type is also used for water lines.

Concrete pipe for sanitary sewers, storm sewers, and culverts are not generally pressure pipe unless it is necessary to pump sewage uphill. Because of the large sizes of some of the pipe, reinforcement is necessary. The use of elliptical-shaped pipe in storm sewer and culverts construction is increasing. Greater flow capacity can be achieved with an ellipse or flat bottomed concrete arch pipe. Thus it is possible to have the same flow capacity in a pipe without requiring as deep a trench. When elliptical or arch pipe is used in a culvert, it is not necessary to embed it as deeply.

There are four ways in which precast concrete pipe (both pressure and nonpressure) can be made:

1. *Cast and vibrated.* The concrete mixture is placed in forms and compaction is achieved by vibration on the outside and inside of the forms.
2. *Packerhead pipe.* A packerhead machine packs the concrete mixture down into the form under pressure.
3. *Tamped pipe.* The concrete mix is tamped into the forms by long rods.
4. *Centrifugated pipe.* Forms that revolve at high speed are filled with concrete mix.

Asbestos-cement pipe is manufactured from portland cement, asbestos fibers, and silica flour. The types of materials and the manufacturing process used give asbestos-cement pipe many advantages in water supply, waste water, irrigation, drainage, and other liquid and air conveyance systems. Asbestos-cement pipe can be either of the pressure or nonpressure variety. In water supply systems, transmission costs are held to a minimum because asbestos-cement pipe offers little flow resistance (Fig. 3-10).

Another basic type of concrete pipe is cast-in-place. Especially large dimensioned or unusually shaped pipe is often cast in place using one of several techniques. Forms may be built and the concrete placed as in other cast-in-place work; or a rounded trench can be dug to form the

Figure 3-10   Asbestos cement pressure pipe.

bottom of the pipe, a form placed to create the top of the pipe and a plastic "sausage" inflated to form the inside of the pipe. A slip-form machine has also been developed to mold concrete pipe in the trench. These methods are all successful but have rather limited application.

Municipal Water Supply

Today, many of the large cities in the United States and other countries must transport their water supplies long distances in closed conduits. Concrete pipe is an excellent means of conveying potable water under the required pressure. Large-diameter pipe carry the prime supply to concrete reservoirs where it is held until needed. Pumping stations direct the water under pressure to various mains keeping water pressure constant throughout the city. Many pressure valves are needed to keep the flow constant; these are usually found in concrete valve vaults at intervals along the mains.

Some cities are fortunate to be located on a body of fresh water that, with proper treatment, can be used as a safe and constant source of drinking water. Filtration and treatment plants are almost always concrete structures.

Rural Water Supply

In rural areas each household is usually responsible for its own water supply. Well casings, well houses and platforms, pump houses, storage tanks, and cisterns are all made of concrete.

In pastures where ground seepage occurs, it is possible to collect this water using concrete tile, to direct it into underground collection boxes (where dirt and sand can settle out), and to watering troughs for farm animals. All parts of this type of spring box are best constructed of concrete.

Waste Water

Concrete is an ideal material for the construction of waste water treatment works. Properly designed and constructed structures have certain inherent advantages: durability, watertightness, strength, beauty and versatility.

Sewers. Sewers carrying the waste water to the treatment plant are usually made of concrete. Concrete pipe is economical for sewer construction because it can be made with local labor, using locally produced materials or materials produced within reasonable hauling distance. Loss from breakage is minimized. Using precast units, sewers can be constructed rapidly, expertly, and inexpensively. Concrete can be molded and finished smoothly. This produces a minimum of frictional drag to the flow of sewage, and allows for the use of smaller pipe. Its long life and low maintenance costs means low annual cost.

A material's durability is established by the way it satisfactorily resists the damaging effects of service conditions to which it is subjected. For sewers, these conditions are (1) weathering, (2) possible chemical action, and (3) wear. The performance of precast concrete pipe and cast-in-place sewers everywhere in the country, under almost every conceivable service extreme, is confirmation of concrete's durability as a material for sewer construction. The quality of well-made concrete to resist the action of freezing, thawing, wetting, drying, and temperature variation is well known. Because sewers are generally constructed underground, weathering action is minimized, and the durability of concrete in sewers under such exposure conditions is unquestionable.

Cast-in-place concrete may be designed and built to fit any desired shape of sewer section. A high-arch section may be preferable to a circular or flat arch for greater structural capacity. The shape of the cross-section may be varied to secure desirable hydraulic characteristics, as when a self-cleaning velocity at varying depths of flow is required or turbulence at a junction is to be minimized. Clearances for obstructions of limited rights-of-way may require other than a normal cross-section.

Concrete can be easily precast into various items that help reduce the cost of sewer construction. Sewer fittings and equipment (elbows, T's, Y's, manhole risers and tops, and catch basins) are easily precast for easier and better construction.

Concrete pipe has been used by the mining industry to carry a mixture of water and highly abrasive mine tailings from mills to tailing ponds. The percentage of solids varied from 4 to 50% and velocities ranged from 2.5 to over 10 ft per sec. Actual performance records are substantiated by tests which indicate concrete will resist the erosive action of clear water at extremely high velocities if there is no abrupt change in direction or velocity of flow that would cause turbulence. These tests indicate that concrete's resistance to abrasion increases as its strength and density increase. Thus, wear is seldom, if ever, a problem in concrete sewers.

Concrete pipe is also used in the construction of storm sewers. Most populated areas require storm sewers to dispose of the runoff from heavy rains or melting snows. In the past, storm sewers were sometimes connected to sanitary sewers, but this practice has been eliminated. Storm sewers usually collect excess runoff, transport it away from the city, and deposit it in a natural waterway.

Septic Systems. For rural residents and city dwellers alike, a modern sewage disposal system is essential. Most rural communities do not have central sanitary systems. Individual homeowners must therefore make adequate provisions for the purification and disposal of the household waste water; this is usually done with a septic tank and disposal field. A typical system consists of (1) a house drain, (2) house sewer line, (3) septic tank, (4) outlet line, (5) distribution box, and (6) properly placed tile in a disposal field.

The septic tank is connected to the house drain by a line of pipe called the house sewer. It is usually built of 6 in. bell-and-spigot concrete sewer pipe. All joints are completely filled with a mortar composed of 1 part portland cement, 3 parts mortar sand, and enough water to provide plasticity. When properly constructed, such joints will normally keep out roots. As a further precaution against root penetration, a mortar band 1 in. thick and 3 in. wide is sometimes placed around the joint.

Septic tanks must be built of a durable material like concrete because they are constantly exposed to moisture and waste products. Precast concrete septic tanks are available in many rural communities.

WATER RESTRAINT

Flood-control measures must be undertaken to minimize flood damage. These involve clearing and maintaining river channels to ensure the passage of excessive runoff without obstruction, constructing dikes and levees to confine the flow in water courses, protecting banks by revetments and floodwalls, and constructing flood detention reservoirs.

Dams

Dams to create flood-control reservoirs are usually built of concrete, earth, or rock, although some older dams still in use are made of masonry or timber.

Selection of a particular type for a given location is usually based on economic considerations. These require a careful evaluation of the primary purpose of the dam, conditions prevailing at the dam site, hydraulic factors imposed by the hydrology and stream's hydraulic characteristics, and climatic conditions.

Many factors influence the design of a dam. Foundation problems, methods of flood control during construction, and construction techniques are decisive factors. The necessity of insuring spillway capacity, spillway-energy dissipation, power generation, and site adaptability must be considered. The methods of solving these and other problems often emerge as national trends. This is evident not only in this country, with its gravity dams, but in countries such as France, where great emphasis has been placed on thin arch dams, and Italy, where there are large number of buttress dams.

Engineers concerned with the design and construction of earthfill dams have spent much time and effort ensuring protection of embankments from the persistent attack of wind and waves. Many materials with considerable range in both cost and durability have been used. None, however, have the slope protection qualities of soil-cement. This mixture of soil, portland cement, and water, when compacted, has the required strength and durability to completely stabilize a slope.

As a construction material, concrete has the advantages of easy placement, control, economy, a relative abundance of raw materials, high compressive and sheer strengths, durability, and low permeability. Depending on the topography and foundation conditions at a particular site, one (or more) of the following types of concrete dams would be suitable: concrete gravity, concrete arch, concrete buttress, and prestressed concrete.

Concrete gravity dams, using their weight, resist force imposed on them. Some are curved in plan; thereby, a portion of the load is carried by arch action. Hoover Dam on the Colorado River and Shasta Dam in California are typical of this type of design and construction. One of the highest and most massive strength dams is the Grand Coulee Dam on the Columbia River in the state of Washington.

Concrete-arch dams are stable because a large portion of the water and other horizontal loads are transmitted into adjacent canyon walls by direct thrust. Examples of this type of dam are Buffalo Bill Dam near Cody, Wyoming, built of rubble concrete shortly after the turn of the century, and Pelton Dam, near Madras, Oregon, constructed in 1958 of conventional concrete.

An arch dam is suitable and economical in a relatively narrow, deep canyon where rock foundations capable of resisting arch thrusts without undue deformation are available.

The design of arch or curved gravity dams once required much more time than the design of straight gravity dams. Recent developments in the adaptation of electronic computers to solve such problems have reduced that time considerably.

Concrete buttress dams are "hollow" concrete dams composed of two principal structural elements: (1) the upstream, water-supporting deck or face and (2) the buttresses that support the deck. The advantages of buttress dams are twofold: smaller quantities of concrete are required than for gravity dams, and unit pressures on the foundation are usually lower.

Stony Gorge Dam in California and Possum Kingdom Dam in Texas are excellent examples, respectively, of early and recent slab-and-buttress dams.

The multiple-arch dam is similar to the slab-and-buttress dam except that the upstream face consists of a series of arch-barrel segments instead of flat slabs. Bartlett Dam in Arizona and Mountain Dell Dam near Salt Lake City, Utah are good examples of multiple-arch dams.

Prestressed concrete dams have been built in limited numbers and only in Europe. This type of design has some advantages and should be given more widespread consideration. Prestressing wires, cables, or bars, anchored in the foundation rock below the dam and extending vertically to its crest, make possible the elimination of some mass concrete required for the stability of this unusual type of concrete dam. Allt-no-Lairige Dam in Scotland, constructed in the mid-1950s, is an example of this type of construction.

Related Concrete Structures

Spillways are designed to prevent overtopping and possible failure of a dam by releasing surplus water from the reservoir. Several types of spill-ways, with their particular characteristics, are described (Fig. 3-11):

1. Overflow spillway. A spillway in which water flows over a concrete dam or the concrete spillway section of an earth dam.
2. Chute spillway. An open channel structure for passing water around a dam into the river channel downstream of it. This type of spillway is most commonly used with earth dams.

Figure 3–11 The overflow spillway on the Grand Coulee Dam on the Columbia Basin Project, Washington. USBR photo by E. Hertzog.

3. Tunnel spillway. This spillway is a tunnel through an abutment of a dam. A tunnel spillway if often used when diversion of a stream during construction of a dam is through a tunnel.

Outlet works for concrete gravity dams and for earth dams with mass concrete overflow spillways usually consist of sluiceways through the concrete. Tunnels or cut-and-cover conduits also serve this purpose adequately. Final choice of type to be used depends on site conditions and economics.

Channel improvements—such as stream straightening and deepening, cutoffs, debris removal, and levees and floodwalls—are necessary to provide complete flood control for a river basin. Floodways extend the benefits of channel improvements further by allowing predetermined areas to act as "safety-valve" reservoirs or channels.

Revetment, or protection of the banks, may be required for cutoffs as well as for natural channels. This work is usually done in durable concrete. Where stream banks or levees are subject to scour, at bends in a river or when current velocities are high, some form of revetment or bank protection is necessary. Many materials—stone riprap, soil-cement, precast concrete in various forms, and concrete paving—have been used successfully for bank protection of levees and natural channels wherever it has been possible to carry on such construction.

The more difficult task of underwater stabilization has been successfully accomplished on a large scale by using articulated mattresses composed of precast concrete sections joined with metal clips and steel cables. Such mattresses have a long record of satisfactory service on the lower Mississippi River.

The flood wall is a special form of levee used in locations such as industrial, residential, or transportation centers where standard levees are not economically feasible. One type of floodwall is the retaining wall sometimes used to provide sidewalls for depressed flood channels.

It is often desirable to completely stabilize the bed of a river or stream. This is accomplished by paving the channel bottom. On large jobs, slip-form pavers are used for the placing of concrete.

Where sufficient quantities of large rock are readily available, grouted cobblestone has been proven satisfactory and economical for invert and slope paving. A stone blanket, usually 12 to 18 in. thick, is constructed of cobbles ranging in size from 5 to 12 in. In a typical operation, the cobbles are placed on a prepared subgrade with clamshell buckets and spread into place with bulldozers. After they have been placed, they are flushed with water to wash down the fines and are covered with a grout of cement, sand, and water.

Occasionally, it is necessary to completely enclose a stream for land reclamation or aesthetic reasons. The top of a closed conduit may serve to support the roadway of a street or alley, frequently the case in congested metropolitan areas. Such conduits are usually designed as rectangular concrete box culverts, sometimes with one or more intermediate walls.

Conduits are concrete pipe or cast-in-place concrete used as overflow mechanisms in floodwater-retarding structures (often dams) to take care of runoff from major storms.

Shore protection, protection of our coastal and inland shorelines against turbulent waters, has been a problem for more than 75 years.

All shoreline erosion is basically caused by two natural actions of the water: incoming waves and littoral currents (those near the shore of an ocean or lake). Studies of many instances of erosion and of the failure of structures intended to prevent erosion have revealed the complexity and great variability of these natural forces. The engineer must draw extensively on the sciences of oceanography, meteorology, fluid mechanics, soil mechanics, structural design, and geology to learn about these forces and how to deal with them.

Shore-protection structures fall into three general classifications: offshore structures, beach-protection structures, and onshore structures. Each has a specific purpose to accomplish and each is most durable when constructed from concrete.

Offshore breakwaters are structures protecting a harbor, anchorage, or basin from waves. They are free standing, located in varying depths of water, and are usually exposed to unobstructed wave action. They may be required in certain instances to protect or maintain the toe of a beach or to trap littoral materials.

Where rock in adequate quantity or size is not economically available, large precast concrete block of various shapes, most commonly cubes and tetrahedrons, have been used. Recently, a shape known as a "tetrapod" has been patented for use in breakwater and jetty construction. The tetrapod resembles a child's jack and consists of a central body from which four truncated cylindrical legs radiate at 120° angles to each other. When in place, the legs of adjoining tetrapods interlock and present a very rough surface that induces wave runup and reflections; large openings between the interlocking legs prevent hydrostatic back pressures (Fig. 3-12).

Figure 3–12   Concrete tetrapods, in foreground, provide breakwater protection for Rincon Offshore Drilling Island near Ventura, California. Photo by ARCO.

Beach-protection structures protect land behind them by absorbing the effects of the waves. They are the most effective means of dissipating wave energy. Beaches are made up of materials eroded by the back shore, brought in from deeper water, or supplied by rivers and streams. These materials are constantly moving, being carried along the beach by littoral currents, or being carried seaward or shoreward by the action of the waves. Whenever an analysis reveals that sufficient littoral materials are available, the use of groins may restore or maintain a beach.

Groins are shore-protection structures, usually constructed perpendicular to the shoreline, for building or widening a beach by trapping littoral drift or retarding loss of beach materials. They are relatively narrow and may extend from less than 100 ft to several hundred feet into the water from a point well landward of any possible shoreline recession.

Jetties differ from groins in that they are much longer and much more massive. They are placed in the entrances of harbors to protect the channels used by ships and at the mouths of rivers to assist in the maintenance of a channel discharging river flows. Like groins, jetties have been built of precast and cast-in-place concrete. Concrete tetrapods have been used successfully and economically in jetty construction at Casablanca, Morocco, to protect a seawater inlet.

Onshore structures are placed approximately parallel to the shoreline and include seawalls, bulkheads, and revetments. Seawalls are massive structures placed to protect upland areas from violent wave action. Bulkheads are ordinarily constructed of lighter materials because their primary function is to retain a fill. Revetments are concrete facings that protect a shore or beach against erosion. There is a wide variety in the shapes of the exposed faces of the walls.

Concrete that is properly proportioned, mixed, and placed is one of the most durable materials available for shore protection. Concrete structures in service all over the world attest to concrete's satisfactory maintenance-free performance.

IRRIGATION

Since America's modern venture into irrigation began, some 27 million acres have been upgraded by a constant water supply. Of these, 25 million acres are in the 17 western states classed as arid or semiarid, areas where rainfall clearly is insufficient to support dryland farming. In other parts of the United States, where rainfall is more abundant during the year, irrigation is receiving more and more attention as insurance against crop failure from lack of moisture at critical times. Arkansas, Louisiana, and Florida each have over half a million irrigated acres, and the areas under irrigation are increasing every year.

A typical irrigation system consists of mainlines and laterals, both of which may be open canals, pipelines, or a combination of these. Many distribution systems are built entirely of underground concrete or asbestos-cement pipe. An irrigation distribution system is built to convey water from a supply canal or reservoir, through mainlines and laterals, to individual ranch delivery points located on these lines. Supplying the required quantity of water at the desired operating level is easily and economically done using a pipe distribution system. Different techniques for regulating these deliveries are required for the three principal types of pipe distribution systems.

Open systems (limited-pressure systems) are the most common type. Vertical open-top stands, equipped with overflow weirs or baffles, are placed in the line to raise upstream line pressures to desired values. The open irrigation system is operated very much like an irrigation system that consists wholly of surface canals and ditches. Both systems utilize gravity to move water and in both, the flow into the laterals and sub-laterals and to the delivery points is controlled by simple slide gates. To avoid wasting water, it is necessary that the amount of water turned into the system be equal to the sum of all the deliveries from it. If more than the desired amount of water is delivered to the system, the excess will be lost through a wasteway or by overflowing one or more of the baffle stands.

Full-pressure systems are similar in principle to municipal water systems. In both cases it is necessary merely to open a delivery valve to get a desired flow of water. Such systems usually are more expensive to construct because they require pressure pipe; however, several factors offset the high initial cost of pipe for the full-pressure system. The most important is that the full-pressure system eliminates waste of water and the necessity for drainage at the end of laterals. The size of the pipe required for a full-pressure system is also smaller than that of the open system.

"Semiclosed" pipe irrigation systems combine the most desirable qualities of the open and full-pressure systems. Inexpensive, unreinforced concrete pipe can be used because the internal pressures in the lines are limited by constant-head float valves. Operating characteristics of the semiclosed and full-pressure systems are quite similar. The semiclosed system seems to be the most economical for lines where the flow and the heads to be regulated by the valves are small. As the semiclosed system is limited in quantity of flow by the maximum size of available constant-head valves, the area served by a lateral is limited. Therefore, the semiclosed system is most suitable where the terrain justifies short laterals from the main service canal or line.

All irrigation systems do not utilize underground pipe. Many have open canals or ditches transporting water from the source to the fields. These open ditches, when unlined, are extremely wasteful.

The earliest irrigation canals were merely unlined ditches, and even today, the great majority of all irrigation canals are unlined. The number of acres that can be irrigated is determined by the amount of water available. Because only water applied to the land contributes to crop raising, losses in transit reduce irrigable acreage. Records of the Bureau of Reclamation and independent irrigation districts indicate that, on the average, almost 40% of the water entering a distribution system of unlined canals never reaches the farm ditches. This includes losses from evaporation, water drawn up by uncontrolled vegetation in and near the canals, unavoidable spilling and waste of excess water, and seepage. The latter is the greatest loss of all, and in canals and laterals averages about 25% of the total water diverted. Losses in some systems run as high as 60%. In some areas, the return flow of seepage water to natural drainage channels or its collection through drains makes its re-use possible, but in most cases it is permanently lost.

When canals, laterals, and ditches are lined with concrete, available water may be distributed as easily and economically as planned. Irrigation district managers and individual farmers utilize every possible means to cut their expenses. Concrete lined irrigation canals and ditches require almost no attention. No weeds need be removed since none grow; concrete will not wear away, scour, or erode; and concrete does not require time-consuming repairs.

Checks in an irrigation ditch make it possible to raise (or check) the water level in a section of ditch for diversion into adjacent fields by setting flash boards in wall grooves in the side of the ditch. Head gates are essential irrigation structures. They divert water from a main canal into individual farm laterals and ditches. They are often equipped with measuring devices that determine the amount of water individual farmers purchase. Where water is obtained from wells and where pumps are used, concrete-lined ditches reduce pumping time. As a result, pumping bills can be cut up to 50% because water moves faster to delivery and less is lost.

If it is necessary to run irrigation water under a road, buried concrete pipe, called inverted siphons, transfer the flow. They act as culverts, carrying irrigation flow beneath the road and out to an adjoining ditch. Unnecessary detours in moving farm equipment are eliminated.

It may be necessary to build an elevated ditch for carrying water across a depression, draw, or swale, or to carry it along a steep hillside or rocky area. This is done in a concrete flume.

Shotcrete (pneumatically applied portland cement mortar) has been widely used for lining and resurfacing irrigation and drainage canals and ditches. Because no great amount of construction equipment is required, this process is well-suited to repair work and small or widely scattered lining projects. Because forms are not needed, shotcrete linings are particularly suited to ditches and laterals where curves, turnouts, and gate outlets are frequent. Shotcrete has been used to line canals of all conceivable shapes and to rehabilitate old concrete-lined canals which retain some strength and structural value. A ½-in. layer of shotcrete is usually enough to provide a smooth, uniform surface that will add many years of service to the original lining.

Accurate measuring assures equitable water distribution to users and provides good control of the application of irrigation water to their crops. For many years the simple concrete weir has been used to measure and regulate flows in both open-ditch and concrete pipe systems. Because of simplicity of design it is probably the most economical measuring device in use. For additional benefit, sand-trapping facilities have been incorporated into many weir boxes. Under some conditions—as where water supplies are limited—weir measurements may not be sufficiently accurate or dependable, and mechanical measuring and totalizing meters have been developed for use in such cases. The vertical flow meter set on top of a riser pipe surrounded by a large standpipe is one form of totalizing meter for ranch delivery units. Where the hydraulic gradient below the meter is too great for an economical flow meter installation, line meters have been used. Regulation of the flow is usually provided by an irrigation-type gate valve installed in the pipe leading to the meter stand.

Whenever water is brought into a pipe system from a reservoir or canal, floating debris must be removed if pipelines are to operate properly. Screens may be installed either in the turnout structure or in a special screening box just below it. When excessive quantities of sand or silt are expected to be in the water entering a pipe system, a sand trap should be provided at the turnout from the canal. Sand traps may be any type of structure that will slow the water sufficiently to permit transported material to settle out, and be readily cleanable.

A collection box should be provided at the upper end of a lateral pipeline when its turnout from the mainline or canal has more than one line of pipe. If the turnout flow is to be divided among several pipe laterals or deliveries, a division box replaces the collection box. A division box is also used wherever several laterals, each too large to be served by a tree, take off from the mainline.

In irrigation pipelines an air vent should be provided wherever air has a chance to accumulate—at high points in the line, at breaks in grade, or downstream from a gate or valve where vacuum may be high enough to overload the gate leaf or cause cavitation damage. This structure is usually made of precast concrete or asbestos-cement pipe.

DRAINAGE

Drainage of wet land permits earlier land preparation, speeds seed germination, and, because of improved soil aeration, increases crop production. The cost of a drainage system is often repaid in increased yield in the first few years. Drainage is also necessary in many irrigated areas to remove excess water and prevent accumulation of plant-killing salts. Concrete-drain tiles are made in all areas of the country requiring land drainage; they are manufactured with inside diameters ranging from 4 to 24 in.

Concrete is used in the construction of several other structures designed to improve soil drainage. Weir notch dams are used to eliminate soil erosion by dropping water from a higher elevation to a lower one. They are also used as checks in waterways. The size of the weir notch dam depends on the amount of rainfall, the area to be drained, the dominant type of vegetation, and the terrain. These structures can be built of either cast-in-place concrete or concrete masonry.

Concrete drop boxes are often used on the inlet side of culverts to reduce erosion. The water level in the ditch rises to the top of the box before water enters the culvert. The box has a stilling action and some settlement of silt and water-borne earth occurs. Some farmers have used drop boxes successfully to build up the earth in eroded ditches as well as to prevent erosion.

A drop inlet, often called a tube and riser, is used to control the water level of dammed farm ponds. It consists of a vertical riser built to the desired water height and extending downward through the dam to a horizontal conduit. Riser and conduit are either reinforced concrete pipe or cast-in-place concrete. The riser must be adequate to accommodate large amounts of water.

Farmers are not the only people who need properly drained soil. Highways constructed over well-drained subbases last longer and give better service. Proper design and construction of drainage facilities helps provide effective and safe airfields. Runways or landing strips remain safe for use if surface and subsurface waters are handled effectively. Ponding of water on runways can immobilize an airport; subsurface water can create pavement failures through frost upheavals and weakened sub-base soil strength.

RECREATION

One outcome of the progress in industry and the sciences since the beginning of the century has been the development of mechanization to the point where man does not have to work as hard or as long as he once did. The laborer who put in 12 to 14 hr of hard work for six or seven days a week in the late 1800s has descendants who work 8 hr a day, five days a week. Indications are that this will tend to decrease further. The industrial revolution, a period of social and labor reform, has resulted in a recreational bonanza.

Outdoor living and the weekly cookout have become a way of life in many localities, and this casual type of outdoor entertaining and living has become very common. Concrete has been able to contribute to the comforts of outdoor life. Patios are often constructed of cast-in-place concrete (sometimes with an exposed-aggregate finish) or of precast concrete units. Concrete tennis courts provide a fast and smooth surface. Swimming and wading pools as well as bath houses are usually constructed of concrete. Shuffleboard courts, children's play areas, and horseshoe pits are just a few of the home recreational facilities made of concrete.

Away from home, concrete provides for reacreational activities in the form of bandshells, stadia, grandstands, public and private boat-launching ramps and docks, and other facilities built for fun and enjoyment. The use of concrete as an art form is also becoming more common.

MISCELLANEOUS USES

Portland cement paint has white portland cement as its principal ingredient. Sold as a dry powder in many colors, the consistency of rich cream is obtained when water is added. It is used on concrete block, stucco, plaster, concrete, and is suitable for both interior and exterior surfaces.

Concrete storm cellars are built in areas where severe storms occur. Asbestos cement is used to make roofing and siding materials as well as wall panels used in industrial building, home construction, and remodeling. Concrete burial vaults are required by law in most states.

Concrete masonry is used to seal off old mine tunnels to direct available ventilating air to areas where the miners are working. Stoppings, as these seals are called, are also used to seal off natural gas leaks or to smother fires.

Peterborough, Ontario, is justly proud of its new fountain. This fountain, centennial symbol of the town, jets water 250 ft into the air. Its unusual feature is not that the platform supporting it is made of concrete —29 tons of it—but that it floats. Filled with foamed polyurethane plastic, the concrete base is buoyant enough to stay afloat summer and winter; the problem of possible ice damage in the cold months is eliminated.

There are other uses for concrete that floats: boats. The ribs of a boat are covered with a wire reinforcing mesh and plastic concrete is placed on the mesh. The concrete boat is as buoyant as a wooden one and has the advantage of not requiring costly upkeep.

One innovative builder in Dallas, Texas, has developed a method of house construction that is cheap, strong, quick, and attractive. Burlap bags, filled with a mixture of sand, gravel, and cement, are piled on top of each other, reinforced with steel bars, and dampened. The concrete mixture hydrates and sets, forming both the exterior and interior walls of the house. The walls are sprayed inside and out with a stuccolike coating, and, with the completion of the roof and floor, the house is ready for occupancy. The walls are thick enough to eliminate the need for insulating material, and the overall, adobelike appearance of the house is very pleasant.

Another builder in Fort Lauderdale, Florida, has developed a cast-in-place concrete roof for use on homes.

Concrete has also been used to repair sheet-steel piling. Concrete reinforcing from 6 in. below the normal water level to high water is poured around a piling previously enclosed in an air-tight chamber and pumped dry. The piling is thoroughly cleaned, and a reinforcing mesh is attached. A form to hold to the concrete is built, the concrete is placed, and it is allowed to cure. Using this method, it is possible to add many years to the life of the original piling.

The Soviets have recently completed construction of the world's tallest free-standing tower. It rises 1750 ft into the air from a prestressed concrete base. Its main function is a TV sending station, but there are four viewing platforms located at various heights and a rotating restaurant at 1050 ft.

The use of soil-cement as low-cost material for housing has proved satisfactory in some areas. Homes of rammed, cast-in-place soil-cement have been built in Mexico and in some parts of Africa.

SUMMARY

A significant trend can be noted from this chapter on the uses of concrete. The pace of development of various types of concrete construction is increasing. Many of the major developments in concrete are relatively recent. More and more use of prestressed concrete members in all types of construction is anticipated. The precasting of dwelling units is so new that there is not much literature on the techniques used, yet these units are thought by many designers, architects, and city planners to be the buildings of the future.

SECTION TWO
MATERIALS
FOR CONCRETE

CHAPTER 4
PORTLAND CEMENT

COMPOSITION OF PORTLAND CEMENT

Portland cement is a finely ground material consisting mainly of compounds of lime, silica, alumina, iron, and gypsum. When mixed with water it forms a paste that hardens and binds other materials, called aggregates, together. The final product is a hard, strong mass called concrete.

Each of the major constituents of portland cement has an abbreviation used by the cement industry that does not correspond to its counterpart in chemical terminology. (See Table 4-A.)

The chemical reactions that take place when water is added to cement are extremely complex and are not even completely understood by chemists. What is known, however, is that by combining the four basic ingredients (oxides of calcium, silicon, iron, and aluminum) with a fifth (gypsum, to control the setting time), a product is created that is plastic and workable for a few hours, becomes firm to allow finishing, and then rapidly becomes strong and durable. By varying the proportions of these five ingredients during manufacture, the properties of cement can be varied to produce fast or slow hardening, resistance to soil chemicals, and other special qualities. When these materials are heated during processing, water and carbon dioxide are driven off and new compounds are formed.

Table 4–A

Compound	Chemical Symbol	Cement Industry Abbreviation[a]	Percentage in Cement
Lime	CaO	C	60–66
Alumina	Al_2O_3	A	3–8
Silica	SiO_2	S	19–25
Iron	Fe_2O_3	F	1–5

[a] In chemistry; C means Carbon, S is Sulphur, and F is Fluorine.

Table 4–B

Cement Industry Abbreviation	Compound	Chemical Symbol	Function
C_3S	Tricalcium silicate	$3CaO \cdot SiO_2$	Cementing compound
C_2S	Dicalcium silicate	$2CaO \cdot SiO_2$	Cementing compound
C_3A	Tricalcium aluminate	$3CaO \cdot Al_2O_3$	Flux during manufacture
C_4AF	Tetracalcium aluminoferrite	$4CaO; Al_2O_3 \cdot Fe_2O_3$	Flux during manufacture

The product of this process is then ground to a fine powder called portland cement. It consists of four basic compounds, each of which has a name, chemical symbol, and a cement-industry abbreviation (Table 4-B). C_3S contributes to early and later strengths, and C_2S gives strength at later ages. C_3A contributes to early strength but is susceptible to attack by sulfates. C_4AF contributes little to the early strength of concrete but does add to later age strength. It is also susceptible to damage by sulfate attack.

The percentage of each compound can be approximated by chemical analysis. Table 4-C shows typical compound composition and fineness for each of the principal types of portland cement.

When water is added to cement at the ready-mix plant or on the job, the two materials combine chemically. The series of complex reactions that take place are collectively called hydration. The process begins rapidly, giving off heat, then it slows down, and, if water for hydration is present, continues for many years until hydration is complete and the cement paste reaches its maximum strength. The higher the fineness, the

Table 4–C
TYPICAL CALCULATED COMPOUND
COMPOSITION AND FINENESS OF PORTLAND CEMENT

Type of Portland Cement		Compound Composition, Percent[a]				Fineness, sq cm per g[b]
ASTM	CSA	C_3S	C_2S	C_3A	C_4AF	
I	Normal	50	24	11	8	1800
II		42	33	5	13	1800
III	High early strength	60	13	9	8	2600
IV		26	50	5	12	1900
V	Sulfate resisting	40	40	4	9	1900

[a] The compound compositions shown are typical. Deviations from these values do not indicate unsatisfactory performance. For specification limits see ASTM C150 or CSA A5.
[b] Fineness as determined by Wagner turbidimeter test. (Approximate Wagner values are obtained by dividing by 1.8.)

greater the rate at which cement hydrates, the earlier strength is developed, and the more rapidly heat is generated.

MANUFACTURE OF PORTLAND CEMENT

Raw materials (Fig. 4-1) provide the lime, silica, iron, and alumina used in making portland cement. Natural deposits of limestone and shell beds are common sources of lime. Clay, shale, slate deposits, and blast furnace slag supply silica, alumina, and iron. Some impure limestones or marl can supply all four basic ingredients, though not always in the right proportions for a desired portland cement composition; these natural combinations are called "cement rocks."

Raw materials may contain too much or not enough of one or more of the essential ingredients; in such cases, materials of suitable composition are added to adjust the raw mix to the desired proportions.

The grayish color of cement is caused by iron compounds and manganese. White portland cement is produced in some mills by using specially selected raw materials so that the end product has less than 0.50% iron and less than 0.02% manganese.

Natural deposits used as raw materials for making cement may contain magnesia, alkalis, and phosphates. These cause detrimental chemical reactions in concrete if present in appreciable quantity in the finished cement. Additions may have to be made or some special processing may have to be done to limit such material. In a cement plant, raw materials are first subjected to preliminary treatment (crushing, screening, etc.) and are stockpiled (Fig. 4-2). Based on chemical analyses, the propor-

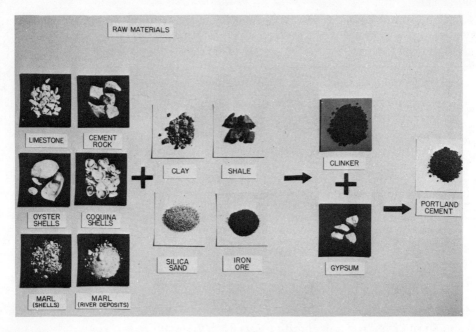

RAW MATERIALS

LIMESTONE   CEMENT ROCK   CLAY   SHALE   CLINKER

OYSTER SHELLS   COQUINA SHELLS   SILICA SAND   IRON ORE   GYPSUM   PORTLAND CEMENT

MARL (SHELLS)   MARL (RIVER DEPOSITS)

Figure 4–1

tions of the raw materials that would result in a finished cement of desired composition are calculated. A "raw mix" is prepared by grinding and blending the chosen materials, sometimes in a dry form and sometimes in the form of a slurry (Fig. 4-3). The wet process (using slurry) is favored by many modern cement plants because of the relative ease of blending. A finely ground, intimate mixture of the raw materials is essential to attain as complete a reaction as possible during burning.

The prepared mixture is fed into the upper end of a rotary kiln (Fig. 4-4), a long cylindrical steel furnace lined with fire brick and capable of rotating slowly on its slightly inclined axis. Modern commercial cement kilns vary from 300 to 700 ft in length and 12 to 25 ft in diameter. As the kiln rotates, the materials pass slowly from the upper end to the lower; the rate is controlled by the kiln's slope and speed of rotation. This process is continuous, as opposed to the batch process used in the vertical kilns of early cement making. The kiln's fire may be fueled by gas, oil, or pulverized coal fed into the kiln at its lower end. As raw cement material passes through the kiln, water and carbon dioxide are driven off, and the temperature is raised to a critical level at which the essential reactions can take place (Fig. 4-6).

By a controlled rate of travel through the kiln, the material is held within this critical temperature range for a period of time sufficient to allow a combination of the lime-silica-alumina-iron mixture to occur, forming the four characteristic compounds of portland cement. (See Fig. 4-1.) Depending on the raw materials to be processed, maximum temperatures developed near the lower end of the kiln generally range from 2600° to 3000°F.

The process of heating the ingredients to produce the desired product is known as "burning" or "calcining." The output of the kiln after this burning process is fused or semifused, rough-textured lumps or pellets ranging in size from $\frac{1}{16}$ to 2 in. in diameter. The substance is called "clinker."

Various reactions take place as the materials move from the upper end down through the kiln. It has been estimated that 20 to 30% of the mix is liquefied in the hotter part of the kiln. The major reactions that produce the characteristic clinker compounds take place here. Temperatures that are too high, however, may induce the formation of undesired compounds and may also result in severe manufacturing difficulties. A balance must be struck to include the rate of reaction, fuel costs, and deterioration of kiln and lining due to high temperatures.

After burning, the clinker is cooled and stored, a step that is sometimes accelerated by spraying with water.

After crushing, the clinker is ground in a ball mill with small amounts of a calcium sulfate material, usually gypsum, until the desired degree of fineness (Fig. 4-5) is achieved. The principal function of the gypsum is to control the cement's time of set when it is mixed with water on the job. Small amounts of other materials are sometimes added during the grinding process, either to facilitate grinding or to impart special properties such as air entrainment to the finished cement.

TYPES OF PORTLAND CEMENT

There are five types of portland cement currently in use in the United States. Requirements for these cements are given in the American Society for Testing and Materials Specification C150.

Type I is the most commonly used cement and is available from all mills. Type II often provides somewhat greater resistance to disintegration by aggressive chemicals, notably the sulfates, found in some soils and water; however, this type of cement may gain strength more slowly and have a lower heat of hydration than Type I. The high-early-strength

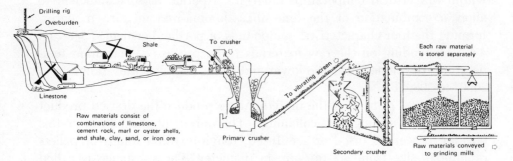

Figure 4–2 This flow chart shows steps in the manufacture of portland cement. The stone is first reduced to 5-in. size, then ¾-in., and stored.

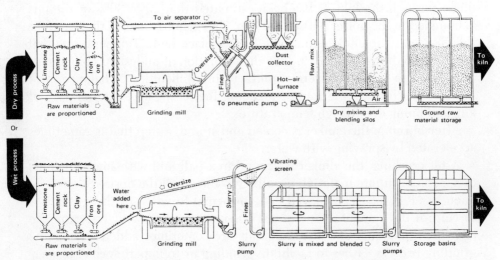

Figure 4–3 Raw materials are ground to powder and blended. Then, the raw materials are ground, mixed with water to form slurry, and blended.

cement, Type III, gains strength more rapidly than the others and is used in cold weather concreting where high early strengths are needed, as well as in most prestressed and some precast structural members. It is generally readily available. Type IV cement has relatively low generation of heat during hydration, and Type V cement has much greater resistance to the action of sulfates in soils or water than the other cements. Because of their rather special nature, Types IV and V are not usually carried in stock and must be made on special order.

In addition to the five basic types of portland cement, three variations are produced by incorporating in them, during grinding, chemical additives which cause the entrainment of air. The air-entraining cements are designated as Types IA, IIA, and IIIA, and correspond to Types I, II, and III; requirements for these cements are given in ASTM Specification C175. The air-entraining cements are being used more widely where there is expectation that the concrete will be exposed to severe frost

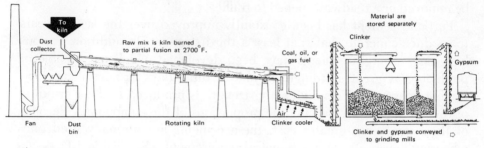

Figure 4-4   Burning changes the raw mix chemically into cement clinker.

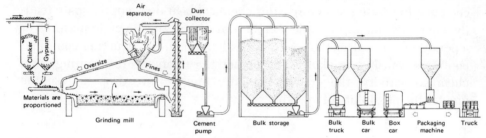

Figure 4-5   Clinker with gypsum added is ground into portland cement and shipped.

Figure 4-6   Modern kilns vary in length from 300 to 700 ft in length and 12 to 25 ft in diameter.

action. Air entrainment can also be obtained by adding a suitable chemical to the concrete during the mixing process.

SPECIAL CEMENTS

High-Early-Strength Cements

On many construction jobs, extra strength is needed quickly so forms can be removed or a roadway opened to traffic.

Portland cement has been constantly improved over the last 50 years to produce concrete that is at least a third stronger. In addition, various high-early-strength cements have been developed which are capable of producing as great a strength in one to three days as normal portland cement does in 28. This high early strength is produced by finer grinding of the clinker and by varying the proportions of the compounds. Because of their somewhat greater cost, these cements are normally used only when high early strength is a special requirement. (See Table 4-D.)

The heat of hydration of such cements is higher than that of normal cements. In massive structures, this chemical heat does not dissipate readily and causes an appreciable temperature rise in the concrete mass. Over a period of time, after the concrete has hardened, the structure cools and contracts, causing cracking. The higher the temperature resulting from chemical activity, the greater the later temperature drop and so the greater the possibility of cracking.

Table 4-D
CURRENT TYPES OF PORTLAND CEMENT

General Description	Designation	Use
General purpose or normal	Type I	For general concrete construction where special properties are not required
Modified general purpose	Type II	For general concrete construction exposed to moderate sulfate action or where it is required that the heat of hydration be somewhat lower than for normal cement
High early strength	Type III	For use when higher early strengths are required
Low heat	Type IV	For use where it is required that the heat of hydration be a practicable minimum
Sulfate resisting	Type V	For use where a high resistance to the action of sulfates is required

Portland-Pozzolan Cements

Pozzolans, siliceous-aluminous materials that react with calcium hydroxide in the presence of moisture, are used to reduce the expansion in concrete caused by chemical reactions between cement and some aggregates and when it is important to reduce internal temperatures in massive concrete structures. Together with the free lime produced in cement hydration, they can form additional cementitious compounds. Pumicite and diatomaceous earth, for example, have pozzolanic properties. A number of the portland-pozzolan cements that have been commercially produced in recent years contain about 25% pozzolan, although other percentages are possible. Compared to portland cements with similar characteristics, portland-pozzolan cements have lower early strength, but final strengths may be greater. Extended periods of moisture availability or high temperature steam curing are necessary to develop the benefits of pozzolanic action.

Specifications for portland-pozzolan cement are given by ASTM C595. Type I cement having appreciable pozzolanic additions is designated as Type IP. If it also contains an air-entraining agent it is designated as Type IPA.

Pozzolans are often added to the concrete batch separately. This procedure eliminates stocking a special type of cement, but requires an extra batching hopper.

Portland Blast-Furnace Slag Cement

Portland blast-furnace slag cement is an intergrind of portland cement clinker and granulated blast-furnace slag with the addition of a small amount of gypsum to control set. If ground with a Type I clinker, it is designated as Type IS. The early strengths obtained with this cement are almost as high as for a Type I portland cement, and at 28 days the strength may be even higher. Its cost is a little lower.

Slag Cements

Slag cements are produced by intergrinding suitable blast-furnace slag with hydrated lime. Since strengths obtained with slag cements are generally low and variable, their use is limited to unimportant structures or to those massive structures where strength is not of great importance. Some slag cements are used as masonry cements.

Masonry Cements

Masonry mortars, used in brick and concrete block work, require masonry cements having greater plasticity than that obtainable with ordinary portland cements. These cements consist of blends of cement and other materials. The cementing materials may be portland cement, portland-pozzolan cement, natural cement, slag cement, hydraulic lime, or hydrated lime. Materials that may be blended with the cementing materials are limestone dust, chalk, talc, pozzolans, and clay. Some of these act as plasticizers to increase the plastic qualities of a masonry mortar.

ASTM Specification C91 covers two types of masonry cement: Type I for general-purpose, nonload-bearing masonry, and Type II for masonry where higher strength is required.

Expansive or Shrinkage-Compensating Cements—Self-Stressing Cements

The chemical composition of these products includes a hydrate of calcium sulfoaluminate for expansive separation. Because of the inherent tendency of ordinary portland cement concretes to shrink on drying, with a consequent tendency to crack when restrained from shrinking, many efforts have been made to eliminate or reduce drying shrinkage. One approach is to prepare a mixture containing compounds that will expand sufficiently to offset the shrinkage tendency of normal hydrated cement compounds. Interest in the possibility of "expansive" cements has been augmented by the increased use of prestressed concrete. It would be desirable in some prestressed applications to utilize the expansive properties of concrete to induce prestress in the steel.

Expansive cement compensates for shrinkage by expanding rapidly at first, with 60 to 80% of the expansion occurring in the first 24 to 36 hours. The total expansion is usually completed in four to seven days. The expansion causes a mild prestressing of the reinforcing steel and provides a slight compressive stress in the concrete. Use of reinforcing steel with expansive cements is essential in reducing cracking; steel percentages of $\frac{1}{2}$ to 3% are most effective. Existing external objects such as walls or slabs should not be depended on to produce the same restraining effect because they lack resiliency. Similarly, covers of burlap or polyethylene sheet membrane as well as water may be used for curing, but expansions will be higher if excess water is supplied. Although all aspects of expansive cements have not been explored, investigation so far shows that when used with general good concreting practices, expansive cement can effectively reduce cracking due to drying shrinkage.

High-Alumina Cement

Although not of the same composition as portland cement, aluminous (or high-alumina) cement is a hydraulic cement that is used to make concrete in much the same manner as regular portland cement. It has certain useful special properties, such as rapid hardening and high resistance to the action of sulfate waters, if used at a relatively low water-cement ratio (0.4 by weight). However, because of its high heat of hydration, aluminous cement is undesirable for use in massive structures such as dams. With special aggregates, such as crushed firebrick, it can be used to make refractory concrete capable of withstanding high furnace temperatures.

White Portland Cement

White portland cement is a true portland and is manufactured to conform to the specifications of ASTM C150 and C175. The principal difference between white and gray cement is their color. White portland cement is made of selected raw materials containing negligible amounts of iron and manganese; the manufacturing process is controlled so that the finished product will be white instead of gray. It is used primarily for architectural purposes such as precast curtain wall and facing panels, terrazzo surfaces, stucco, cement paint, tile grout, and decorative concrete. Its use is recommended wherever white or colored concrete or mortar is desired.

Oil-Well Cement

Oil-well cement is used to seal oil well walls. It must not flash-set when subjected to high temperatures and pressures. The American Petroleum Institute Specification for Oil-Well Cements (API Standard 10A) covers requirements for six classes of cement. Each class is applicable for use at a certain range of well depths. The petroleum industry also uses conventional types of portland cement together with suitable set-modifying admixtures.

"Waterproofed" Portland Cement

"Waterproofed" portland cement, which is actually water-repellent cement, is usually produced by adding a small amount of calcium, aluminum, or other stearate to portland cement clinker during final grinding. It is manufactured in either white or gray color.

Plastic Cements

Plastic cements are made by adding plasticizing and air-entraining agents to Types I or II cement during the manufacturing process. They are commonly used for making mortar, plaster, and stucco.

A number of physical characteristics of cement are important to the specifier. These include fineness, time of set, soundness, strength, heat generation, and sulfate resistance. In some cases, characteristics such as heat of hydration, alkali content, false set, and air content of mortar made with the cement may be of special interest. Tests for all these requirements must usually be performed in a well-equipped laboratory. Standard tests and limits of acceptability for these properties are set by ASTM (American Society for Testing & Materials), AASHTO (American Association of State Highway and Transportation Officials), and the federal government. With the help of these standards, the user may specify, for instance, cement that meets ASTM specifications for Type I. He will then know that the product must fall within certain limits of prescribed chemical and physical requirements. If the user has reason to believe that the cement he has purchased does not meet his requirements, he may have it checked for compliance with the standard by an independent laboratory. If the user is having trouble with the time of set, he may want to have the fineness and the time of setting evaluated. It is usually unnecessary to have cement tested for all its properties.

FINENESS

The rate of reaction of cement with water is greatly influenced by the size of the cement particles. For a given weight of a finely ground cement, the surface area of the particles is greater than that of coarsely ground cement. This increases the rate of reaction with water and speeds the hardening process. If, however, a cement is too finely ground, the extremely fine particles may be prehydrated by moisture present as vapor in the grinding mills or during storage. No cementing value can be derived from such particles. Furthermore, the completeness with which a cement can react with water is influenced by the particle size. The cores of coarser particles may take years to hydrate under practical conditions, and there is evidence that very coarse particles may never completely hydrate.

The fineness of cement is currently stated in terms of specific surface, which is the calculated surface area of the particles, in square centimeters per gram of cement. This is commonly measured either by the Blaine air-permeability test or the Wagner turbidimeter test.

While the calculated specific surface is only an approximation of the true surface area, it serves very satisfactorily as a measure of fineness;

good correlations are obtained between specific surface and the properties of cement influenced by fineness of particles.

TIME OF SET

Concrete—a mixture of cement, water, and aggregates—must remain in a plastic condition long enough to permit transporting, placing, and compacting under a variety of practical, and sometimes adverse, conditions. It is also desirable for economy and other reasons that the mass should harden and develop strength within a reasonable time after it has been placed. Understanding the setting and hardening process is essential to quality control in the field.

Although fresh concrete can be readily manipulated for periods of 1 to 2 or more hours after mixing, it gradually stiffens. As hydration proceeds, not only do reaction products take up what was originally "free" water, but the gel and other products begin to occupy more space, decreasing the mobility of the paste. Finally, increasing numbers of gel particles and other products make sufficiently close contact to develop bonds of strength; if the mass is left undisturbed, it begins to develop rigidity. At some point, the mass can sustain a more or less arbitrary load without flowing, and the paste is said to have set.

The water content of a paste has a marked effect upon the time of set. In acceptance tests of cement, the water content is regulated by bringing the paste to a standard condition of wetness, called "normal consistency" —that condition for which the penetration of a standard weighted needle into a paste is 1 cm in 30 sec.

In order to avoid the complications of having a mineral aggregate present, tests for time of set are performed on a "neat" cement paste, that is, a cement and water mixture. The time of set of neat paste at normal consistency is determined by the ability of a specimen of the paste to sustain the weight of specified small rods or needles. These determine "initial" set and "final" set.

Initial set indicates the beginning of noticeable stiffening. Final set is taken to indicate the beginning of what may be regarded as the hardening period in pastes of this consistency.

The standard specifications for cements of Types I through V require that initial set shall not occur in less than 1 hr as determined by the Gillmore needle (or 45 min by the Vicat needle), and that final set shall occur within 10 hr. Most portland cements meet these requirements by a wide margin. In general, they produce initial set in 2 to 4 hr. and final set in 5 to 8 hr. In normal concretes, with considerably higher water cement ratios, set occurs later than in neat paste of standard consistency.

The setting process of a normal cement paste proceeds through four distinct stages. The first stage, which lasts several minutes after the initial contact of cement with water, produces a relatively high rate of heat generation. The cement grains are wetted and initial dissolution of the compounds and reaction of their constituents begins to take place; thereafter, the rate of heat generation drops to a relatively low value. The second stage, sometimes called the "dormant" period, lasts for 1 to 4 hr. During this period, activity proceeds at a low rate. Cement grains slowly build up an initial coating of reaction products, and sedimentation (or bleeding) and loss of slump take place. The third stage begins as heat evolution rises again. In this stage, the degrees of rigidity described as initial and final set are reached. As the activity of the third stage subsides, the fourth stage, the period of hardening, begins.

In a cement having insufficient gypsum, an extremely rapid reaction (within a few minutes) results in a "flash set." Flash or quick set is the rapid development of rigidity in a freshly mixed cement paste or concrete, usually accompanied by the evolution of considerable heat. This rigidity cannot be overcome and plasticity cannot be regained by further mixing unless water is added.

Another phenomenon known as "false set" sometimes occurs. False set is the rapid development of rigidity in a cement paste, but without much heat generation. It is generally caused by cement that has been subjected to high temperatures during grinding or storage. This results in the dehydration of gypsum and produces plaster of paris. Rigidity can be overcome and plasticity regained by further mixing without the addition of water. Continued mixing or agitation breaks up the crystalline mass that has formed and the corrected mixture can be used without any bad effects.

Rapid setting is not the same as rapid hardening. A rapid setting cement is one in which the reactions take place at such a rate that the mass becomes rigid and begins to acquire some degree of strength at earlier ages than normally retarded portland cement. Too-rapid setting of cements intended for normal use may cause difficulties in handling. Conversely, there may be conditions that dictate cement setting very rapidly. To stop the flow of water in tunneling operation joints, special, very rapid-setting cements may be used. To obtain setting in a reasonable length of time under cold-weather conditions where the normal setting process would be unduly delayed, accelerators are sometimes added to the mixture. A rapid-hardening cement is one in which the reactions and strength gains take place at a greater rate than in an ordinary, general-purpose (Type I) cement. Accelerated hardening can be obtained not only by using a cement with rapid-hardening properties but by using admixtures that accelerate the reactions by inducing temperatures that favor more rapid hardening or by using richer mixtures.

SOUNDNESS

Unsoundness in cement is caused by expansion of some of the constituents; the result is cracking, disruption, and disintegration of the concrete. One source of unsoundness in cement is the hydration of free lime encased within the cement particles. The protective cement film prevents the immediate hydration of the free lime. However, moisture may eventually reach the lime after the cement has set, and since lime expands with considerable force when hydrated under restraint, its delayed hydration may disrupt the concrete. Cements manufactured in compliance with ASTM, AASHTO, and federal specifications are free from this type of unsoundness. If the danger of such a reaction exists, a slow-setting concrete may ease the problem by allowing the lime to hydrate before the mass becomes rigid.

Another possible cause of unsoundness is the presence of too much magnesia. Standard specifications limit the magnesium-oxide content to 5%.

Fine grinding of raw materials brings them into closer contact when burned, eliminating the chance of free lime existing in the clinker. Thorough burning of the raw materials further reduces the amount of free lime. Finally, fine grinding of the clinker tends to expose free lime so it can hydrate quickly before the entire mass of concrete hardens.

One method for detecting unsound cements is the "autoclave" test (ASTM C151). An autoclave is a high-pressure steam boiler. Test specimens—bars made of neat cement paste of normal consistency, 1×1 in. and $11\frac{1}{4}$-in. long—are placed in the autoclave 24 hr after molding and subjected to a high-pressure steam atmosphere, 295 psi, for 3 hr. After cooling, the length of the bars is measured and compared with their original length. Portland cements that exhibit an expansion of not more than 0.8% are considered sound. The high temperatures used in these tests accelerate the hydration of the cement to develop conditions that would require much longer periods under normal curing.

STRENGTH

The strength of cement is usually determined by tests on mortars made with it. Both tension and compression testing have been standardized by ASTM.

Tensile-strength tests are made on mortar briquettes having a net cross-sectional area of 1 sq in. and composed of 1 part cement to 3 parts sand, by weight.

In the standard compressive-strength test, the specimen is a 2-in. mortar cube with a composition of 1 part cement to 2.75 parts sand, by weight. Required strengths vary with curing time and run from 1000 to 3000 psi. Most modern cements easily exceed these values.

Table 4-E shows how the relative compressive strength of concrete at various ages is influenced by the compound composition and fineness of the cement.

HEAT OF HYDRATION

The setting and hardening processes of concrete are caused by the chemical reaction of cement and water. Like most chemical reactions, hydration of cement is accompanied by heat release. Concrete is a good insulator. In a large mass of concrete such as a dam, the heat generated during setting and hardening is not readily dissipated. This results in a rise in temperature and an expansion of mass; later cooling and contraction of the hardened concrete could result in development of serious cracks.

In massive concrete structures, temperature rise can be controlled by

Table 4–E
APPROXIMATE RELATIVE STRENGTH OF
CONCRETE AS AFFECTED BY TYPE OF CEMENT

Type of Portland Cement		Compressive Strength, Percent of Strength of Type I or Normal Portland Cement Concrete			
ASTM	CSA	1 day	7 days	28 days	3 months
I	Normal	100	100	100	100
II		75	85	90	100
III	High early strength	190	120	110	100
IV		55	55	75	100
V	Sulfate resisting	65	75	85	100

slowing down the rate of placement, by substituting ice for part of the mix water, by cooling the aggregates, and, in some large structures such as Hoover Dam, by a system of embedded cooling pipes. The amount of heat generated can be materially reduced by modifying the cement composition. This is accomplished by limiting the amounts of tricalcium silicate and tricalcium aluminate.

The heat of hydration of normal portland cement is about 85 to 100 cal per g; for a low-heat cement, such as the type used in the construction of Hoover Dam, this figure is 60 to 70 cal per g. Various procedures have been used by different investigators to determine the heat of hydration. The ASTM method is based on measurement of the heat of solution in hydrofluoric acid of hydrated samples of the cement.

STORING AND SHIPPING OF PORTLAND CEMENT

Portland cement may be stored indefinitely without losing its cementing qualities if it is protected from moisture. For this reason it should be stored in as dry and airtight a location as possible. Bulk cement should be stored in weatherproof bins designed to prevent dead storage in corners. If bagged cement must be stored out of doors, the sacks should not be placed directly on the ground and should be covered for protection against rain. When bagged cement is stored for a long period of time it can become stiff or lumpy. This condition (called "warehouse pack") can usually be corrected by rolling the sack on the floor. If the cement still contains lumps that cannot easily be broken up, it is advisable not to use it.

Cement is shipped in bulk by rail or truck to ready-mix concrete producers and block manufacturers or directly to construction sites. It is quickly unloaded by automatic equpment into silos or hoppers for storage. Cement is also sold in paper bags. In the United States, a bag holds approximately 1 cu ft of cement (loose measure) and weighs 94 lb.

CHAPTER 5
WATER

INTRODUCTION

Water initiates the chemical reactions that produce the binding qualities of portland cement. Without it, cement is merely so much unusable powder, and concrete is impossible to make. Just as there are standards for acceptable cement, so too are there standards for acceptable water.

EFFECTS OF IMPURE MIXING WATER

In order to hydrate, cement must have the proper amount of proper water in the paste. It is important to know the source of this mixing water. If the water has not been tried previously, or has made unsatisfactory concrete, there are a number of points to consider and a variety of tests that can be run.

Impurities concentrated in sufficient quantity in water could seriously reduce the strength of concrete and cause wide variations in setting time. Certain other chemical compounds may cause unattractive discoloration in concrete surfaces, efflorescence (movement of salts to the surface), and excessive corrosion of reinforcing steel.

Water that has the effect of reducing the quality of concrete to a point below minimum acceptable standards allowable for a particular application should not be used.

TESTING WATER FOR MIXING

If the water to be used for mixing is in any way questionable, it should be submitted for lab analysis before using. There are two ways of evaluating mixing water. One practical method is to prepare mortar cubes using the questionable water, and compare the 7- and 28-day compressive strengths of these cubes with others made using water known to be satisfactory. If the strength of the questionable cubes is at least 90% of the strength of the other cubes, the water is judged to be acceptable.

The second method of evaluating the water is to have it analyzed for its chemical content. If the mix water in question is not from a municipal water supply, an analysis is more difficult to obtain. However, a municipal water-department laboratory, a university, a private laboratory, or a local chemical company may be willing to perform the analysis. The results can be checked against a table of standards (Table 5-A) to ascertain whether or not the water falls within acceptable limits.

Analysis of a city's water supply is always available from the municipal water department (Table 5-B). It will list dissolved solids (silica, iron, sodium, and sulfate) in the water as so many milligrams per liter (ppm*). Water containing less than 2000 mg per liter of total dissolved solids can generally be used satisfactorily for making concrete. Although higher concentrations are not always harmful they may affect certain cements adversely and should be avoided where possible. A good rule of thumb to follow is if water is pure enough for drinking, it is suitable for mixing concrete.

IMPURITIES IN MIXING WATER

Sea water is suitable for mixing concrete if the concrete is not to be reinforced. Sea water has a maximum concentration of 3.5% salts in its composition. It makes concrete of higher than normal early strength, but this strength is reduced at later ages. Compensation for strength reduction may be made by reducing the water-cement ratio. In reinforced concrete, increased corrosion potential is another effect. This risk can be reduced if the reinforcement is given sufficient concrete coverage (3 in.),

---

* Parts per million.

Table 5–A
SUBSTANCES IN CONCRETE MIXING WATER

	Maximum Content (ppm or ml/1)
Salts	
Sodium carbonate and bicarbonate	1,000
Calcium and magnesium carbonates	400
Magnesium sulfate and chloride	40,000
Sodium chloride	20,000
Sodium sulfate	10,000
Acids	10,000
Iron salts	40,000
Silt or suspended particles	2,000
Sea water	35,000
Industrial Wastes	4,000
Sanitary Sewage	400
Sugar	500
Algae	1,000
Potassium and sodium hydroxide	0.5-1.0% (by weight of cement)
Oils	2.0% (by weight of cement)

Table 5–B
TYPICAL ANALYSES OF CITY WATER SUPPLIES
(parts per million)

Analysis No.	1	2	3	4	5	6
Silica (SiO_2)	2.4	0.0	6.5	9.4	22.0	3.0
Iron (Fe)	0.1	0.0	0.0	0.2	0.1	0.0
Calcium (Ca)	5.8	15.3	29.5	96.0	3.0	1.3
Magnesium (Mg)	1.4	5.5	7.6	27.0	2.4	0.3
Sodium (Na)	1.7	16.1	2.3	183.0	215.0	1.4
Potassium (K)	0.7	0.0	1.6	18.0	9.8	0.2
Bicarbonate (HCO_3)	14.0	35.8	122.0	334.0	549.0	4.1
Sulfate (SO_4)	9.7	59.9	5.3	121.0	11.0	2.6
Chloride (Cl)	2.0	3.0	1.4	280.0	22.0	1.0
Nitrate (NO_3)	0.5	0.0	1.6	0.2	0.5	0.0
Total dissolved solids	31.0	250.0	125.0	983.0	564.0	19.0

if the concrete is watertight, and if it contains adequate amounts of entrained air.

Sea water increases surface dampness and efflorescence. For this reason, it should not be used where surface appearance is a major consideration or where concrete is to be painted, plastered, or otherwise decorated.

Sea water should never be used in making prestressed concrete. Data collected from laboratory tests and field situations have shown that chlorides present in sea water greatly accelerate prestressed steel corrosion.

Sand and gravel extracted from sea water are sometimes used as concrete aggregates. The amount of salt remaining in the aggregate is usually not more than 1% the weight of the mixing water. Such aggregate, when used with suitable water, contains less salt than sea water used for mixing. Some aggregates react unfavorably with sea water, expanding abnormally.

Although sea water as a mix water must be used with care, it can produce satisfactory results. It was used in the concrete mix for the foundation of a lighthouse built by the U.S. Corps of Engineers in 1910 at the far end of the Los Angeles, California, breakwater. Twenty-five years later the concrete was examined and found to be in good condition, with sharp-edged corners and no disintegration. Much of the concrete used in construction of the Florida East Railway causeway (now a highway to Key West) was mixed using sea water with no detrimental effect.

Sodium chloride and sulfates appear in relatively large amounts in natural waters found in the western United States. Heavy concentrations of these compounds—up to 20,000 mg per liter (2%)—are generally acceptable. Mixing waters containing 10,000 mg per liter (1%) of sodium sulfate have been used satisfactorily.

Carbonates and bicarbonates, while uncommon in significant amounts in natural water, have significant effects on the setting times of different cements. Sodium carbonate may cause very rapid setting; bicarbonates may either accelerate or retard set. These salts can materially reduce concrete strength. When the sum of these dissolved salts exceeds 1,000 mg per liter, tests for setting time and 28-day strength should be made.

Miscellaneous inorganic salts may be present in mixing water. If it is suspected that significant amounts of salts are dissolved in the water, it is good policy to test the concrete for retarded setting time or reduced strength. In general, concentrations of salts up to 2000 mg per liter can be tolerated in mixing water.

Acid or alkaline water used in mixing concrete does not usually affect concrete's strength. The main exception is certain mine water containing high concentrations of sulfuric acid.

Sewage and industrial substances in mix water—sugars, citrates, various types of acids, oils, sewage, and peat fibers—can delay the setting of concrete or lower its strength. These materials are found in waters carrying sanitary sewage, in industrial waste waters, or in waters from bogs or stagnant ponds. In many cases, this type of water produces satisfactory concrete. However, such mixing water should not be used without prior testing.

Waters carrying sanitary sewage may contain about 400 mg of organic matter per liter. After the sewage is diluted in a proper disposal system, the concentration is reduced to about 20 mg per liter or less. This amount is low enough not to have any significant effect on concrete strength.

Tannic acid, humus, or peat fiber may be present in bog water. Such water will have a dark color. It should not be used as a mixing water unless tests have shown that it will not significantly reduce strength.

Water containing sewage or industrial waste usually has less than 4000 mg of total solids per liter. When such water is used in mixing concrete, the reduction in compressive strength is generally not greater than 10%. However, waste water from paper pulp mills, tanneries, paint factories, coke plants, and chemical and galvanizing industries may contain harmful impurities. It is best to test any waste water containing even a few hundred milligrams of unusual solids per liter.

Sugar is one organic compound that causes considerable concern. Small amounts (0.03 to 0.15%) will generally retard setting time; larger amounts (over 0.20%) may accelerate it. There have been actual cases where concrete did not set because the aggregates had been packed in unwashed sugar barrels or because unwashed candy or molasses buckets had been used to measure mixing water. The careless tossing of a soft drink on wet concrete can prevent surface hardening. Less than 500 mg per liter of sugar in mix water will usually have no adverse effect on the concrete, but if the concentration exceeds this amount, tests for setting time and concrete strength should be made.

Oils of various kinds are occasionally present in mix water. Generally such water makes poor concrete. Mineral oil, not mixed with animal or vegetable oils, probably has less effect on strength development than do other oils. However, even mineral oil in concentrations greater than 2% by weight of cement may reduce the concrete strength by more than 20%.

Silt in mixing water is permissible to a tolerance level of 2000 mg per liter. Higher concentrations may not affect strength, but could influence other properties of concrete. Muddy water should remain in settling basins to allow silt to settle out before use.

Algae are a simple form of plant life found in bogs and other bodies of stagnant water. They give water a bright green color. Algae in water or on the aggregate can appreciably reduce concrete strength by interfering with the bond between aggregate and cement or by causing large amounts of air to be unintentionally entrained in the concrete.

CHAPTER 6
AGGREGATES

INTRODUCTION

Aggregates for concrete are mainly sand and gravel; they are used as filler or bulk to extend concrete volume. It is essential to know something about aggregates in order to be able to select those that will give concrete the desired properties at the lowest price. This chapter discusses properties of good concrete—durability, strength, and economy—and shows which aggregate characteristics contribute to these qualities. Table 6-A summarizes aggregate characteristics and lists pertinent ASTM and CSA designations.

DURABILITY

Durability is the ability of concrete to withstand freezing and thawing, wetting and drying, heating and cooling, and abrasion and deterioration resulting from chemical reactions.

Resistance to freezing and thawing is essential for durability. Examining a rock under a microscope usually reveals tiny holes and passageways running throughout. When these holes, called pores, become filled with water, the rock is saturated. As water freezes, it expands; if confined where there is no room to expand, as in a pore, the rock holding the

Table 6–A
AGGREGATE CHARACTERISTICS

Characteristic	Significance or Importance	Test or Practice, ASTM and CSA Designation	Specification Requirement
Resistance to abrasion	Index of aggregate quality. Warehouse floors, loading platforms, pavements	C131	Max. percent loss[a]
Resistance to freezing and thawing	Structures subjected to weathering	C290, C291	Max number of cycles
Chemical stability	Strength and durability of all types of structures	C227 (mortar bar) C289 (chemical) C586 (aggregate prism) C295 (petrographic)	Max. expansion of mortar bar[a] Aggregates must not be reactive with cement alkalies[a]
Particle shape and surface texture	Workability of fresh concrete		Max. percent flat and elongated pieces
Grading	Workability of fresh concrete. Economy	C136 A23.2.2	Max. and min. percent passing standard sieves
Bulk unit weight	Mix design calculations. Classification	C29 A23.2.10	Max. or min. unit weight (special concretes)
Specific gravity	Mix design calculations	C127 (coarse aggregate) C128 (fine aggregate) A23.2.6 (fine aggregate)	
Absorption and surface moisture	Control of concrete quality	C70, C127, C128 A23.2.6, A23.2.11	

[a] Aggregates not conforming to specification requirements may be used if service records or performance tests indicate they produce concrete having the desired properties.

water must be strong enough to resist pressure. It must stretch, or it must break. An aggregate's resistance to freezing and thawing (as well as heating and cooling and wetting and drying) is called "soundness."

The soundness of a rock or aggregate depends on a number of factors. These include size, distribution, shape, and amount of pore interconnection, the amount of water that will be absorbed by the rock, the strength of the rock, and the size of the particle. Concrete containing aggregates that cannot resist freezing and thawing is subject to general disintegration. If only a few particles with poor characteristics are present, a phenomenon known as a "pop-out" occurs. This is a situation in which expanding coarse aggregate particles near the surface push off the outer layer of mortar, leaving holes. Some chert, clay and shalelike materials are well known for this behavior. In fact, any aggregate that has a porous chalky appearance should be suspected, as should gravel or stone containing absorptive iron oxides, clay balls, or organic matter. Even air entrainment may not prevent the deterioration of concrete made with unsound aggregates.

Resistance to wetting and drying is subject to the same characteristics that affect resistance to freezing and thawing. Although this problem is not as serious as that of freezing and thawing, in rare instances, the swelling and shrinking caused by moisture gains and losses can be enough to cause failure of the concrete.

Resistance to heating and cooling prevents damage even if the temperature ranges are great enough or the changes are fast enough. Most common aggregates can be used under ordinary temperature conditions without any detrimental effects to the concrete. In theory, an aggregate that expands at a rate different from that of the cement paste surrounding it should produce stresses and strains in the concrete. In rare cases, where the coefficient of aggregate expansion differs greatly from that of the cement paste, serious problems could easily follow.

Abrasion resistance is the ability of concrete to resist wear caused by heavy foot or wheel traffic. This characteristic is probably most important on highways, heavily traveled sidewalks, and industrial floors. A high quality, well-cured cement paste is essential to resisting abrasion; however, the hardness of the aggregate is also an important factor. Aggregates such as trap rock, granite, and quartz, noted for hardness, are often used where high abrasion resistance is required.

Alkali-aggregate reaction is a special problem and is not covered in specifications other than those in areas where this problem is known to exist. Most common is the reaction between alkalis in the cement, or from other sources, and reactive siliceous materials in the aggregate.

The alkali-aggregate reaction may cause abnormal expansion and cracking of concrete. The best information for the selection of nonreactive aggregates can be found in field-service records. If an aggregate is suspected of being chemically unsound and no service records are available, one of three laboratory tests may help. A preliminary step is often an examination by a qualified petrographer. This involves microscopic examination and identification of potentially reactive substances. Mortar-bar tests may also be made, following the ASTM C227 method. The expansion developed in small mortar bars is measured. The objection to using this test is the time factor; often, three to six months elapse before conclusions can be drawn. Another test, ASTM C289, can be completed in three days. In it, the degree of reaction between a sodium hydroxide solution and a crushed specimen of aggregate is measured. This test, however, is not as definitive as the mortar-bar type.

If aggregates suspected of alkali reaction must be used for economy reasons, a "low-alkali" cement should be specified. If low-alkali cement is not available, an appropriate pozzolanic material or a mineral admixture should be used provided it contains no harmful alkalis.

Another reaction known as "reaction of carbonate aggregates" results from certain carbonate particles that have been found to be chemically reactive in concrete.

Based on field observations and laboratory tests, fire resistance in concrete can be achieved by using blast furnace slag, calcareous aggregates, and aggregates of igneous origin. These categories of aggregate include such rock types as limestone, dolomite, granite, obsidian, and others.

Acid resistance is associated more with cement paste than with aggregates. However, acid-resistant aggregates are required for some special uses: concrete exposed to farm silage, fertilizers, animal wastes, and industrial chemicals, and concrete used in breweries, dairies, and paper mills. In cases where exposure to acid is severe, nonacid-resistant aggregates such as limestone or dolomite are often used so mortar and aggregates may erode uniformly. Used in this way, the "sacrificial" aggregates extend the life of the structure.

STRENGTH

Strength in concrete, second in importance only to durability, is usually measured in two ways: by compressive tests or by flexural tests. Compressive strength is concrete's ability to withstand squeezing or crushing forces. Flexural strength is its ability to withstand bending and resulting tension. The strength of concrete usually depends more on the strength of the cement paste and on the bond between it and the aggregate than on the strength of the aggregate alone. This cement-aggregate bond is affected greatly by the texture and the cleanliness of the surface. A rough-textured surface normally bonds better than a smooth one; however, the texture and shape of an aggregate have more effect on fresh than on hardened concrete. Rough or angular particles need more water to produce a workable mixture than do rounded or smooth particles. Flat, elongated particles require even more mixing water. The percentage of these flat shapes in a mix should be limited to about 15% by weight of the total aggregate. To produce workability with rough or angular aggregates, it is necessary to add water; then, to maintain the strength of the concrete, more cement must be added. As a result, concrete with rough or angular particles will have a higher cement content and be more costly than concrete with rounded particles.

Coatings of clay, dust, oil, algae, or chemical substances on aggregate prevent its proper bonding with cement paste. This can ruin an otherwise perfectly satisfactory mix. When a concrete test cylinder made with aggregate that has a dirty surface is crushed, the aggregate can often be picked out of a perfectly smooth bed, showing that foreign matter has prevented bond between cement and aggregate. (A concrete with very low strength may also produce this effect.)

SHRINKAGE

Shrinkage takes place principally in the cement paste as concrete dries. The amount of shrinkage depends on the amount of water added to the mixture, the cement content, the maximum size of the aggregate, its shape, grading, and cleanliness. If concrete with a minimum amount of shrinkage is needed, water content should be kept low.

THERMAL PROPERTIES

Thermal properties of aggregate give concrete the ability to conduct heat. The degree to which it expands when heated is greatly affected by properties of the aggregates. If these qualities must be controlled in the concrete, selection of proper aggregates will be necessary to achieve the desired result. Testing for these properties involves complex laboratory procedures beyond the scope of this course.

MODULUS OF ELASTICITY

Modulus of elasticity is a measure of the way a material supports a load. A high modulus of elasticity means that a heavy load can be applied to the concrete with relatively low deformation. The same factors that increase concrete's strength also increase its modulus of elasticity. Both the type of aggregate and its size and grading have a great effect on this property, just as they do on strength. Hard, flinty aggregates produce concrete with a high modulus of elasticity; soft aggregates produce low values.

SKID RESISTANCE

Skid resistance of concrete is a major concern on roads and runways. Research indicates that, initially, the surface texture of concrete is more important than the aggregate in preventing skidding. However, as the surface of the concrete wears, the importance of the aggregate increases. One characteristic that reduces skid resistance is the tendency of some aggregates to become polished as the concrete surface wears away. The Virginia Council of Highway Investigation and Research investigated the polishing characteristics of aggregates extensively. They found that a large proportion of fine aggregate particles between $\frac{1}{8}$ in. and $\frac{3}{8}$ in. produce a skid-resistant surface. Fine aggregates tend to leave large percentages of the harder minerals.

ECONOMY

Economy in concrete production is reflected in aggregates. With the exception of water, aggregates are usually the cheapest component of concrete, and they make up about 75% of its volume. Therefore, the cost of the aggregates alone is significant. In addition, aggregates greatly affect the amount of cement necessary to produce strength and other concrete properties. The size, shape, texture, and grading also affect the workability and the cost of placing the concrete. Aggregate cost depends on how difficult the aggregate is to obtain, the supply that is available, the processing that must be rendered, and the distance it must be transported to a job site. This cost must be weighed carefully against the cost of the required cement and that of placing and finishing.

In any particular area, only a few basic types of aggregates are generally available. If these perform satisfactorily, they are almost always the most economical to use. Aggregates may be brought in from other areas for special projects, but the resultant cost will be much higher than that of local materials.

COARSE AGGREGATE—TYPES, QUALITIES, AND CLASSIFICATION

The accurate identification of an aggregate and its constituents is usually the task of a trained geologist or petrographer. However, some knowledge of the basic types can often be helpful to the concrete technologist. Even extremely accurate identification can never provide an entirely acceptable basis for predicting the behavior of an aggregate in service. The physical condition of the aggregate at the time of use, the intended use of the concrete, the qualities of the cement matrix with which the aggregate is surrounded—all these factors have a major effect on performance. The problem is further complicated by the fact that most rock is composed of multiple mineral grains. In general, natural aggregates such as crushed stone or crushed gravel can be broadly classified according to their origin as being igneous, sedimentary, or metamorphic. This classification can further be subdivided by the petrographer according to the aggregate's mineral and chemical composition, texture, and structure.

Igneous rocks are those that have been formed by a molten mass cooling; they are either coarse-grained, fine-grained, or glossy. The coarse-grained rocks, such as granite, cooled slowly beneath the earth's surface; fine-grained types, such as pumice and basalt, were formed by emitted molten lavas and frequently contain natural volcanic glass. Sedimentary rocks (e.g., sandstone, chert), as the name implies, were laid down in strata, usually under water. Metamorphic rocks (e.g., marble) are those that, during the evolution of the earth, were subjected to pronounced and usually turbulent changes under heat and pressure. As a result, the structure of metamorphic rock is more compact and crystalline.

Tables 6-B and 6-C outline some of the main characteristics of rocks likely to be considered for aggregates and the minerals of which they are composed. The tables refer to natural materials only; that is, they do not include blast-furnace slag or expanded lightweight aggregates. Naturally, some of the minerals mentioned will be largely eliminated by processing or restricted by specification limits; examples are the clays, mica, and limonite.

PROCESSING

Processing, as well as handling and storing, affect important aggregate properties: gradation, uniformity of moisture content, cleanliness, and, in the case of crushed aggregate, particle shape. Subsequently, these

Table 6–B
THE COMMONEST MINERALS
(IN APPROXIMATELY DECREASING ORDER OF OCCURRENCE)
LIKELY TO BE FOUND IN CONCRETE AGGREGATES

Name	Nature	Characteristics	Appearance
Feldspar	Most abundant rock-forming minerals. Very wide range covering potassium, sodium and calcium types	Softer than, and can be scratched by, quartz	Good cleavage, usually to show several smooth surfaces, frequently covered with fine parallel lines
Quartz	Silica (silicon dioxide)	Hard, will scratch glass, but cannot be scratched by knife	When pure is colorless with a glassy luster and a shell-like fracture. No visible cleavage and, in massive rocks, no characteristic shape
Opal	Hydrous form of silica. Principal constituent of diatomite. Usually found in sedimentary rock. Relatively rare	Amorphous. Specific gravity and hardness less than quartz. Water content from 2 to 10%. (NB reactive with the alkalis in portland cement)	Color variable. Luster from resinous to glassy. No characteristic external shape or internal crystalline arrangement
Chalcedony	Submicroscopic mixture of fibrous quartz with a smaller but variable amount of opal. Relatively rare	Intermediate between quartz and opal. Frequently occurs in chert. Is reactive with the alkalis of portland cement	Distinguishable from quartz and opal only by laboratory test
Tridymite Cristobalite	Crystalline forms of silica. Rare	Reactive with cement alkalis	Sometimes well-shaped crystals, otherwise laboratory identification only
Zeolite	Comprises a large group of soft hydrous silicates	Can be reactive with cement alkalis	Usually white or light colored
Hornblende Augite	Most common of ferromagnesian minerals. Silicates of iron and/or magnesium. Frequently found in marble		Dark green to black
Calcite	Calcium carbonate	Relatively soft and readily scratched by knife (calcite dissolves	Rhombohedral cleavage; that is, they break into fragments
Dolomite	Calcium carbonate		

Table 6-B (*Continued*)

	and magnesium carbonate	with effervescence in cold dilute hydrochloric acid; dolomite only if sample or acid is heated or sample is pulverized)	with parallelogram-shaped sides
Mica		Perfect cleavage; can be split into very thin flakes	Colorless, light green, dark green, dark brown, and black
Pyrite	Sulfide of iron. Less common as marcasite. Known commonly as "fool's gold." Can be both reactive and nonreactive with cement alkalis	Frequently occurs as cubic crystals. Reactive types can be distinguished by immersion in lime water. Reactive (staining) types produce blue-green precipitate within a few minutes that turns brown on standing; nonreactive types are unaffected	Brass yellow in color with metallic luster
Magnetite Hematite Limonite	Iron oxides		Black Reddish Brown or yellowish

(From *Ready-Mixed Concrete*, March 1965.)

properties have an important influence on concrete quality. Cost limits the amount of processing that can be done to achieve desirable properties.

Aggregate processing can be divided into two broad classifications: (1) basic processing—for gradation and cleanliness and (2) beneficiation—to remove deleterious materials.

Basic processes typically employed to provide aggregate of satisfactory gradation and cleanliness include the following:

1. Crushing and grinding—stone, slag, and large gravel require crushing to provide the required distribution of sizes. Crushing (and possibly grinding) is sometimes used to produce sand where none is available. Crushing gravel, however, may produce an excess of finer sizes or undesirable particle shapes; these may adversely affect workability or increase the water requirements of concrete.
2. Screening—the primary process for producing desired gradation in the coarse aggregate size range. Although dry screening of crushed

Table 6–C
THE COMMONEST ROCK TYPES

Name	Nature	Minerals Present
Granite	Medium- to coarse-grained, light-colored igneous rock	Mainly feldspar and quartz, the former usually being the more abundant. Dark-colored mica also usually present and light-colored mica frequently. Hornblende occasionally
Syenite	Same as for granite	Essentially feldspar. Quartz generally absent. Hornblende maybe
Diorite	Structure as for granite but darker in color than either granite or syenite yet lighter than gabbro	Essentially feldspar plus hornblende or other ferromagnesian minerals
Gabbro (often known together with other types under the collective term "trap" or "traprock," which refers to all dark-colored, fine- to medium-grained rocks).	Medium- to coarse-grained, dark-colored igneous rock	Essentially ferromagnesian minerals plus less abundant feldspar. Usually rich in calcium
Obsidian Pumice Perlite Felsite Basalt	Fine-grained igneous rocks containing natural glass. Color normally light but can range to dark red or even black. Felsite and basalt often listed under the term "trap" or "traprock"	A high silica content means that they may be reactive with cement alkalis
Limestones Dolomites	Sedimentary types. Includes very soft chalk (or lime rock) and marl (high clay content). Limestones recrystallized by metamorphism are known as marble	General term, carbonate rocks. If more than 50% of the carbonate constituent is known to consist of the mineral dolomite, limestones are called dolomites. Silica, clay, sulfates, and organic matter are also present
Sandstone Quartzite	Particles of sand or gravel cemented together. If on fracture the rock breaks around the grains it is a sandstone, if through them a quartzite. Sandstones are sedimentary: quartzites may be sedimentary or metamorphosed types	Cementing material may be quartz, opal, calcite, dolomite, iron oxide, clay, or other material

Table 6-C (*Continued*)

Claystone Siltstone Shale Slate	Sedimentary argillaceous rocks. Characterized by a laminated structure and a tendency to break into thin particles. Hardness follows as result of progressive metamorphosis	According to silt and clay from which they were formed
Chert, also, when dense yellow, brown or green, known as jasper. Dense black or grey types frequently known as flint	Very fine-grained sedimentary rock. Very hard (scratches glass but is not scratched by knife blade). Shell-like fracture in dense varieties. Wide color range. Dense types usually grey to black, or white to brown (less frequently green, red, or blue), and with a waxy to greasy luster. Porous varieties (less tough) are lighter in color being white, stained yellowish, brownish or reddish, and with a chalky surface	Silica in the form of chalcedony, quartz, opal, or any combination of these. Chert frequently occurs as nodules or bands in limestones
Gneiss Schist	Formed by metamorphosis of igneous rocks. Characterized by layered structure. Occur in considerable quantities in concrete aggregates. Gneiss is usually coarser grained than schist	Micas in parallel lenses and bands. Quartz and feldspars (more abundant in schists) in granular form

(From *Ready-Mixed Concrete*, March 1965.)

stone and blast-furnace slag is quite common, it may be necessary at times to apply water during the screening process to remove fine particles present in the material. Screening is normally accompanied by the application of wash water, thus expediting the separation of fines from coarse sizes. Usually, screening alone is employed only on sizes larger than the No. 8 sieve, although there are exceptions.

3. Washing—employed to remove silt, clay, and excess fine sand. It usually begins with an application of water during screening and is completed with removal of unwanted fines in the overflow from water classification. If the aggregate contains clay, mud, mudballs, or organic impurities in such quantities or so firmly attached that ordi-

nary washing will not clean it adequately, scrubbing or log washers may be needed.

4. Water classification—sizing and control of gradation in the finer sizes are usually accomplished by classification in water. A wide variety of classifying devices are used for this purpose; all are based on different settling rates of different sized particles. Water classification is not feasible for sizes larger than approximately $\frac{1}{4}$ in. However, gradation can be controlled with considerable accuracy by suitable reblending, in spite of the overlapping in sizes.

"Beneficiation" is a mining-industry term used to describe the improvement in quality of a material resulting from the removal of unwanted types of rocks. Success of this process depends on big differences in the physical properties of desirable and undesirable materials—their hardness, density, and elasticity. The separation method to be used depends on the nature of the individual deposit.

1. Crushing is used to reduce the quantities of soft and crumbly particles in coarse aggregates. Unwanted material is eliminated either by screening or by classification. The costs of installation and operation are likely to be high, and three is always loss of good sound material.
2. High-velocity water and air remove light material, such as wood, miscellaneous trash, and some lignite. A rapidly moving stream of water carries the floatable material, allowing the heavier aggregate to sink. Only large differences in specific gravity will yield efficient separation. In a few instances, high-velocity air has also been employed for this purpose.
3. Hydraulic jigging may be used to separate materials with much smaller differences in specific gravity than those requiring high-velocity water separation. Lightweight shales and cherts are examples of materials that can be removed successfully by this process. A jig is essentially a box with a perforated bottom in which a separating layer is formed by a pulsating water current. The difference in the specific gravities of the materials contained causes a separation between light- and heavyweight particles.
4. Heavy media separation (HMS) entails using a combination of water, finely ground ferrosilicon, and magnetite mixed to a specified specific gravity as a heavy medium for separating various aggregates. The heavier materials sink, and the undesirable, lighter ones float. By closely controlling suspension, minimizing turbulence, and eliminating fine particles (below about No. 8 sieve), accurate elimination of material below a selected specific gravity level can be accomplished.

5. Elastic fractionation is a recently developed process of extremely limited applicability. It involves dropping the aggregate particles to be tested onto a steel plate from which those with higher elasticity bounce farther than the presumably less desirable particles of lower elasticity. A separation is achieved by collecting the portions of differing rebound in separate compartments.
6. Magnetic separation is the removal of iron from crushed blast furnace slag with electromagnets—highly effective when the iron is in metallic form.

HANDLING AND STORING OF AGGREGATES

Much can be done in aggregate handling to increase the likelihood of good performance. Conversely, basically good material may yield inferior results due to abusive handling. The following are some of the recommendations of the American Concrete Institute for handling aggregate.

1. Segregation in coarse aggregate can be minimized if the aggregate is separated into individual sizes for separate batching.
2. Particles smaller than the designated limit for each size should be held to a minimum.
3. Stockpiles should be built in horizontal or gently sloping layers. Conical stockpiles or any unloading procedure involving the dumping of aggregates down sloping sides of piles should be avoided.
4. Trucks and bulldozers should be kept off stockpiles as they cause breakage and contamination.
5. Effective measures should be taken to insure accurate separation of sizes as the aggregate is deposited at the batch plant.
6. Storage bins should have the smallest practicable horizontal cross section; bottoms should be sloped at an angle not less than 50° from the horizontal toward a center outlet. They should be filled by material falling vertically over the outlet.
7. Two sizes of sand cannot be blended satisfactorily by placing them alternately in stockpiles, cars, or trucks. Blending, if required to improve grading, should be done by feeding the different sizes into a common stream on the belt or loader from regulating feeders. Where two or more sizes of sand are employed, it is preferable to batch them separately.
8. Wind should not be permitted to segregate dry sand.
9. Where practical, wet sand should be drained until it reaches a uniform moisture content. Generally, a satisfactory and stable condition can be reached in 48 hr or less. (See Figs. 6–1, 6–2, and 6–3.)

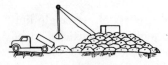

Preferable

Crane or other means of placing material in pile in units not larger than a truckload that remains where placed and does not run down slopes.

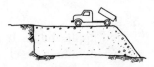

Objectionable

Methods that permit the aggregate to roll down the slope as it is added to the pile or permit hauling equipment to operate over the same level repeatedly.

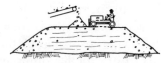

Limited acceptability

Pile built radially in horizontal layers by bulldozer working from materials as dropped from conveyor belt. A rock ladder may be needed in this setup.

Generally objectionable

Bulldozer stacking progressive layers on slope not flatter than a 3:1 ratio. Unless materials strongly resist breakage, these methods are also objectionable.

Figure 6–1 Incorrect methods of stockpiling aggregates cause segregation and breakage.

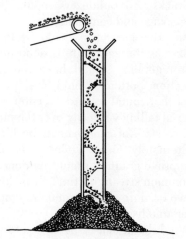

Figure 6–2 Finished coarse aggregate storage. When stockpiling large-sized aggregates from elevated conveyors, breakage is minimized by use of a rock ladder.

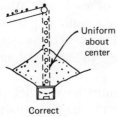

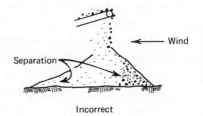

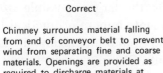

Correct

Chimney surrounds material falling
from end of conveyor belt to prevent
wind from separating fine and coarse
materials. Openings are provided as
required to discharge materials at
various elevations on the pile.

Incorrect

Free fall of material from high end
of stacker permits wind to separate
fine from coarse material.

Figure 6–3   Fine aggregate storage.

METHODS OF DETERMINING AGGREGATE PROPERTIES

Characteristics of finished concrete are influenced by the properties of the aggregate. This knowledge has little practical value, however, without knowledge of how to identify these aggregate properties. There is a wide variety of aggregate tests; some are practical only under laboratory conditions and may require the judgment of a trained petrographer for interpretation. Others are commonly used for specification of concrete material on the job. The equipment for performing these specification tests is available to most aggregate suppliers and concrete plant operators. Some of the tests can be performed directly in the field.

Grading—Particle-Size Distribution

Particle-size distribution, or grading of aggregates, affects not only the water requirement of concrete but such qualities as economy, strength, finishing characteristics, and workability. For this reason, tests of aggregate gradation are very important. In general, aggregates that have a relatively even distribution of particle sizes make the best concrete. When such evenly distributed aggregates are used in concrete, a minimum of spaces or voids remains between particles. Therefore, a minimum of cement paste is needed to fill the voids. This phenomenon can be illustrated by filling a laboratory beaker with 1-in. aggregate particles, a second beaker with ⅜-in. particles, and a third with well-mixed particles of both sizes. By measuring the amount of water it takes to fill each of the three beakers, it can be seen that the beaker with both sizes of particles combined will take less water than the other two beakers. The most common way of measuring the percentage of each size particle in a sample of aggregates is by sieve analysis. The sample is passed through a series of sieves that have decreasing sizes of mesh. The amount of each size particles can be determined by weighing the material retained in each sieve.

Selection of the maximum size aggregate must have is determined by the nature of the job. For economic reasons, the largest possible size is generally preferable, but clearance requirements impose a practical limitation. For mass construction in dams, 4- and even 6-in. maximum aggregate is general practice. For heavy pavements, 2½ in. is acceptable. For structural reinforced concrete, on the other hand, clearance becomes a major factor because of the distance limitations between reinforcing steel bars and between reinforcing and forms. A good general rule here is that the largest pieces of aggregate should not exceed about one-quarter or one-fifth the thickness of the unreinforced concrete or three-quarters the distance between reinforcing bars or between reinforcement and forms. In normal, dense, general-purpose nonreinforced concrete, 1½-in. diameter aggregate has been found to be the most practical for most jobs.

Broad limits for grading requirements of coarse aggregates have been established by the ASTM (Table 6-D). They represent the percentages, by weight, passing a series of standard sieves with square openings.

Specific Gravity

Specific-gravity measurements may be used to calculate the total weights of materials to produce a given volume of concrete. (For instance, how many pounds of aggregates, water, and cement are required per cubic yard of concrete?) The specific gravity of an aggregate is the ratio of its weight to the weight of an equal volume of water. Average specific gravities of some common aggregate materials are: sandstone, 2.50; sand and gravel, 2.65; limestone, 2.65; granite, 2.65; and traprock, 2.90.

Unit Weight

Unit weight (bulk unit weight) of an aggregate is the weight of the amount needed to fill a 1-cu ft container. It is a method of determining the voids or air spaces in a unit volume of aggregate.

Bulking

"Bulking" is a problem caused by surface moisture where the concrete mix is proportioned by volume. Excess moisture present in sand and holding the sand particles apart causes an increase in volume. The amount of

Table 6-D
ASTM GRADING REQUIREMENTS FOR COARSE AGGREGATES
(percent passing, by weight)

Nominal Size	4 in.	3½ in.	3 in.	2½ in.	2 in.	1½ in.	1 in.	¾ in.	½ in.	⅜ in.	No. 4	No. 8	No. 16
3½ in. to 1½ in.	100	90–100	—	25–60	—	0–15	—	0–5	—	—	—	—	—
2½ in. to 1½ in.	—	—	100	90–100	35–70	0–15	—	0–5	—	—	—	—	—
2 in. to No. 4	—	—	—	100	95–100	—	35–70	—	10–30	—	0–5	—	—
1½ in. to No. 4	—	—	—	—	100	95–100	—	35–70	—	10–30	0–5	—	—
1 in. to No. 4	—	—	—	—	—	100	95–100	—	25–60	—	0–10	0–5	—
¾ in. to No. 4	—	—	—	—	—	—	100	90–100	—	20–55	0–10	0–5	—
½ in. to No. 4	—	—	—	—	—	—	—	100	90–100	40–70	0–15	0–5	—
⅜ in. to No. 8	—	—	—	—	—	—	—	—	100	85–100	10–30	0–10	0–5
2 in. to 1 in.	—	—	—	100	90–100	35–70	0–15	0–5	—	—	—	—	—
1½ in. to ¾ in.	—	—	—	—	100	90–100	20–55	0–15	0–5	—	—	—	—

bulking varies with the moisture content and the grading; fine sands tend to bulk more than coarse. Since most sands are delivered in a damp condition, wide variations can occur in batch quantities if proportioning is done by volume. The concrete technologist must know about aggregate moisture and its effect on mix design.

Aggregate Moisture and Mix Design

Aggregate moisture, concrete mix design, and their interrelationship must be well understood. Most natural rock or aggregate particles are porous. They have vacant spaces or pores located between the individual mineral grains comprising the particles.

Pores in an aggregate particle start to fill with "absorbed water" immediately on exposure to wetness. The water absorption rate of a typical limestone aggregate is high when the aggregate is first wetted but rapidly decreases with time (Fig. 6-4).

In the absorption process, air originally filling the pores of a dry aggregate particle must be displaced by water. As absorption continues, intruding pore water encounters increased difficulty in displacing or absorbing shrinking air bubbles located progressively deeper inside the aggregate particle. This is believed to be the cause of continually decreasing absorption noticed in aggregates even years after they are first immersed in water.

A dry aggregate used in a concrete mix will absorb some water from

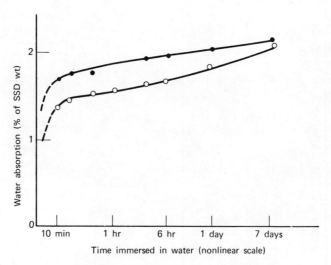

Figure 6-4

the mix. Since aggregate pores are usually too small to admit cement grains, absorption of water by dry aggregate in a mix dehydrates the paste. Only water in intimate contact with paste particles can dilute the paste, lubricate the mix, or regulate mix strength and durability. Because absorbed water is isolated away from the paste in holes in the aggregate, it is not included in calculating the mix's water-cement ratio.

An illustrated description of the terms describing various aggregate moisture conditions together with the changes in mix water-cement ratio caused by their use is given in Figure 6-5. Note that an aggregate is either dry and absorbs water from the paste or, on rare occasions, is saturated surface-dry (SSD) and does not affect paste water content. Occasionally, it is wet and sheds its "free" or "surface" moisture into the paste during mixing, thereby diluting it.

Because it is extremely difficult to keep aggregate moisture content at the saturated surface-dry state to avoid absorption or surface-moisture effects on a concrete mix, concrete producers almost always use aggregates in the condition in which they are received. By determining the moisture condition of samples of aggregate and making the appropriate moisture corrections, adequate control of concrete quality can be maintained as aggregate moisture conditions vary. Procedures for doing this are subsequently outlined.

Moisture Properties of Aggregate Samples. Four weight determinations characterize the moisture conditions of a sample taken from large aggregate stocks. Two are for artificial, laboratory induced conditions:

Oven-dry aggregate weight (D in grams)—the weight of the sample after it has been dried to constant weight in an oven at 105 to 110°C on a hot plate or when heated with burning alcohol.

Saturated surface-dry aggreate weight (SSD in grams)—the weight of initially wet coarse aggregate that has been surface-dried with a towel or of aggregates that have been dried in air to the point of surface dryness.

Two sample weights characterize the materials as they come from the stockpile. Depending on the moisture condition of the aggregate, it will have a weight that is:

Air-dry aggregate weight (A in grams)—the weight of dry-surface coarse aggregates or of noncohesive fine aggregates.

Term	Aggregate Moisture Condition	If Used in Mix

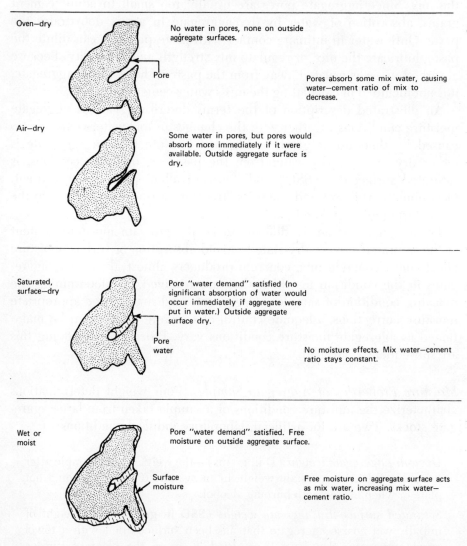

Oven—dry — No water in pores, none on outside aggregate surfaces.

Pore

Pores absorb some mix water, causing water—cement ratio of mix to decrease.

Air—dry — Some water in pores, but pores would absorb more immediately if it were available. Outside aggregate surface is dry.

Saturated, surface—dry — Pore "water demand" satisfied (no significant absorption of water would occur immediately if aggregate were put in water.) Outside aggregate surface dry.

Pore water

No moisture effects. Mix water—cement ratio stays constant.

Wet or moist — Pore "water demand" satisfied. Free moisture on outside aggregate surface.

Surface moisture

Free moisture on aggregate surface acts as mix water, increasing mix water—cement ratio.

Figure 6–5

Wet aggregate weight (W in grams)—the weight of obviously wet coarse aggregates or of cohesive fine aggregates.

The preceding weights are used to calculate the moisture properties of a sample, which are defined as:

Absorption capacity—a measure of the water held in the pores of saturated surface-dry (SSD) aggregates after immersion for a specific time. Absorption capacity (%AC) is expressed as a percentage of the oven-dry aggregate weight (D).

$$\%AC = \frac{\text{wt total pore water}}{D} \times 100$$

$$= \left(\frac{SSD - D}{D} \right) 100 \qquad \text{Eq. 1a}$$

$$D = \frac{SSD}{1 + \dfrac{\%AC}{100}} \qquad \text{Eq. 1b}$$

$$SSD = D \left(1 + \frac{\%AC}{100} \right) \qquad \text{Eq. 1c}$$

Pore moisture—a measure of the water held in the pores of an air-dry aggregate. Pore moisture (%PM) is also expressed as a percentage of the oven-dry aggregate weight.

$$\%PM = \frac{\text{wt water in pores}}{D} \times 100 = \frac{A - D}{D} \times 100 \qquad \text{Eq. 2a}$$

$$A = D \left(1 + \frac{\%PM}{100} \right) \qquad \text{Eq. 2b}$$

$$D = \frac{A}{1 + \dfrac{\%PM}{100}} \qquad \text{Eq. 2c}$$

Surface moisture—a measure of the free or surface moisture on aggregate surfaces. Surface moisture (%SM) is expressed as a percentage of the SSD aggregate weight.

$$\%SM = \frac{\text{wt of surface moisture}}{SSD} \times 100$$

$$= \left(\frac{W - SSD}{SSD} \right) 100 \qquad \text{Eq. 3a}$$

$$W = SSD \left(1 + \frac{\%SM}{100} \right) \qquad \text{Eq. 3b}$$

$$SSD = \frac{W}{1 + \dfrac{\%SM}{100}} \qquad \text{Eq. 3c}$$

Mix Moisture Corrections. Wet aggregate. In a concrete mix, wet aggregate will lose all its surface moisture to the paste, but retain its internally absorbed pore water. Equations 3a, b, and c may be used to relate the original weight of wet aggregate put in the mix and the weight of SSD aggregate to which this weight of wet aggregate becomes equivalent as surface moisture passes into the paste during mixing.

Air-dry aggregate. During mixing, an air-dry aggregate absorbs water from the paste until the aggregate attains a pore water charge equivalent to that held by the same aggregate in the SSD condition. So:

$$\frac{\text{air-dry aggregate wt}}{+\text{wt mix water absorbed}} = \text{SSD wt} \qquad \text{Eq. 4}$$

An approximate* equation for determining the weight of moisture absorbed from the paste in a mix by air-dry aggregate is:

$$\text{wt mix water absorbed} = \text{SSD}\left(\frac{\%\text{AC}}{100} - \frac{\%\text{PM}}{100}\right) \qquad \text{Eq. 5}$$

combining Equations 4 and 5,

$$\text{air-dry wt} = \text{SSD}\left(1 - \frac{\%\text{AC}}{100} + \frac{\%\text{PM}}{100}\right) \qquad \text{Eq. 6}$$

Mix water adjustment. The "net mix water" for a concrete batch is the amount of water in the mix which is actually a part of the paste. In effect, net mix water is all the water in a mix except that absorbed by the aggregates in the mix. The mix water-cement ratio is expressed as a weight-to-weight ratio of the net mix water to the cement in the mix. Net mix water relationships may be expressed as follows:

$$\begin{aligned} \text{net mix water} &= \text{total water in mix-water} \\ &\quad \text{absorbed in aggregate} \qquad\qquad\qquad \text{Eq. 7a} \\ &= \text{water added at mixer} \\ &\quad +\text{aggregate surface moisture} \\ &\quad -\text{wt mix water absorbed} \qquad\qquad \text{Eq. 7b} \end{aligned}$$

Sample Calculations. (Calculating Moisture Properties of Aggregates)

1. Coarse aggregate sample was immersed in water for 24 hr, the saturated aggregate surface-dried with a towel, and its weight determined

---

* The exact equation, calculated by combining Eqs. 1a and 2a, is needed only when using porous aggregates having absorption capacities higher than 5%.

(5101 g). After drying in an oven, the aggregate weighed 5019 g. Find the weight of water absorbed and the percent absorption capacity of the aggregate.

wt absorbed water $= 5101 - 5109 = 82$ g absorbed water

From Eq. 1a,
$$\%AC = \frac{5101 - 5019}{5019} \times 100 = 1.6\% \text{ absorption capacity}$$

2. The original wet weight of a fine aggregate was 519.3 g. When dried to saturated surface-dryness by an alcohol fire, it weighed 502.1 g. Find the weight and the percent surface moisture in the sample.

wt surface moisture $= 519.3 - 502.1$
$$= 17.2 \text{ g surface moisture}$$

from Eq. 3a,
$$\%SM = \frac{519.3 - 502.1}{502.1} \times 100 = 3.4\% \text{ surface moisture}$$

3. The original weight of an air-dry sample of coarse aggregate was 2066 g. When dried to constant weight by burning alcohol, the sample weighed 2051 g. Find the weight of absorbed water in the sample and its percent pore moisture.

wt absorbed water $= 2066 - 2051 = 15$ g absorbed water
$$\% \text{ PM} = \frac{2066 - 2051}{2051} \times 100 = 0.7\% \text{ pore moisture}$$

Calculating Mix Water Corrections. Mix water corrections are of two types: one in which idealized batch quantities (SSD aggregate and net mix water weights) are calculated from trial batch quantities with air-dry or wet aggregates, and the other the correction of day-to-day production batch weights to maintain accurate batch quantities while using aggregates of variable moisture content. Examples 4, 5, and 6 are of the first type; Examples 7, 8, and 9 of the second.

4. A wet sand has 5% surface moisture. What is the SSD weight of 1200 lb of this wet aggregate? How much surface moisture does this weight of wet sand contain?

From Eq. 3c,
$$\text{SSD} = \frac{1200}{1 + 5/100} = 1143 \text{ lb SSD aggregate}$$

From Eq. 3a,
$$\text{wt surface moisture} = W - SSD = 1200 = 1143$$
$$= 57 \text{ lb surface moisture}$$

5. What is the SSD weight equivalent of an air-dry gravel weight of 2100 lb (gravel %PM $= 0.5\%$, %AC $= 1.4\%$)? How much water will this gravel absorb during mixing in concrete?

From Eq. 6 (transposed),

$$\text{SSD} = \frac{2100}{1 - (1.4/100) + (0.5/100)} = \frac{2100}{0.991} = 2110 \text{ lb SSD aggregate}$$

From Eq. 4,

$$\text{wt mix water absorbed} = 2110 = 2100 \text{ lb}$$
$$= 19 \text{ lb mix water absorbed}$$

6. Twelve hundred pounds of the wet sand in Example 4 and 2100 lb of the air-dry gravel in Example 5 were used in a trial mix. If 250 lb of water was added at the mixer, what was the net mix water weight in the trial mix?

From Eq. 7(b) and values in Examples 4 and 5,

$$\text{net mix water} = 250 + 57 - 19 = 288 \text{ lb net mix water}$$

7. Sixteen hundred and seventy pounds of SSD fine aggregate must be used in a mix. Stocks of this aggregate have a surface moisture content of 5.0%. What weight of wet aggregate is equivalent to the required 1670 lb SSD aggregate weight? Calculate the weight of surface moisture carried into the mix by the wet aggregate.

From Eq. 3b,

$$W = 1670 \left(1 + 5/100\right) = 1754 \text{ lb wet aggregate}$$

From Eq. 3a,

$$\text{wt surface moisture} = 1754 - 1670 = 84 \text{ lb surface moisture}$$

8. A coarse aggregate has an absorption capacity of 2%. The aggregate stock has a pore moisture content of 1%. Twenty-five hundred pounds of this coarse aggregate in the SSD condition are required for a mix. How much mix water will be absorbed by this aggregate during mixing? What weight of air-dry aggregate must be batched?

From Eq. 5,

$$\text{wt mix water absorbed} = 2500 \frac{(2-1)}{100}$$
$$= 25 \text{ lb mix water absorbed}$$

From Eq. 4,

$$\text{air-dry wt} = 2500 - 25 = 2475 \text{ lb air-dry aggregate}$$

9. Fine and coarse aggregate having SSD weights and moisture conditions stipulated in Examples 7 and 8 must be batched for a mix together with 346 lb of net mix water. How much water must be added at the mixer to get this net mix water?

From Eq. 7b and values from Examples 7 and 8,

$$\text{water added at mixer} = 346 - 84 + 25$$
$$= 287 \text{ lb water added at mixer}$$

Soundness

The property of soundness usually refers to the ability of an aggregate to withstand weathering (mainly freezing and thawing). The performance of aggregates under exposure to freezing and thawing can be predicted by past performance and by testing. If aggregates from the same source have previously given satisfactory service, the aggregates may be considered suitable. If no such experience record is available, there are two methods of testing: freeze-thaw and sulfate soundness. Laboratory freezing and thawing tests of aggregates in concrete probably provide the best measure of soundness although no standard test has been established for general use. These tests subject air-entrained concrete containing samples of aggregates to alternate cycles of freezing and thawing. Deterioration is measured by changes in the sonic modulus that correspond to changes in strength or by changes in the weight of the length of the samples.

A second method of testing is the sulfate soundness test. This test consists of immersing an aggregate sample in a solution of sodium or magnesium sulfate and then drying it in an oven. This process is repeated several times, causing salt crystals to build up in the pores of the aggregate. This creates an internal pressure.

Poor performance is indicated if a large part of the coarse aggregate will pass a sieve with openings five-sixths of those on which it was retained. Sand is judged to have poor performance if, after test, a large part of the sand will pass a sieve on which it was originally retained. This test has not always been successful in predicting the resistance to freezing and thawing. This is probably because the aggregate is tested in an unconfined state; that is, it is not part of the concrete. Aggregates in concrete are surrounded by cement paste of extremely low porosity; this restricts expansion of aggregate particles and prevents water from saturating particles easily.

A third type of test requires concrete samples containing the aggregate to be soaked continuously in water and periodically subjected to freezing to determine whether critical saturation has been reached. A critically saturated sample will deteriorate when it freezes because its pores are so filled with water that there is no room for freezing expansion; the excess volume cannot readily escape into the air voids contained in the surrounding paste. Concrete that is able to withstand soaking for a period of time equal to the freezing season without becoming critically saturated is considered immune to freeze-thaw action.

Abrasion Resistance
Abrasion resistance measures the wear caused by a combination of impact and surface abrasions. Testing provides an indication of probable break-

age during handling, stockpiling, and mixing and is widely used to indicate overall aggregate quality and strength-producing potential. There are two basic pieces of abrasion resistance testing equipment: the Los Angeles abrasion machine and the Deval machine. Both machines evaluate abrasion resistance from the increase in fines produced by aggregates tumbling with steel balls inside a steel drum. The Los Angeles machine is the more widely used.

Concrete specifications often set an upper limit on weight loss permitted during testing. However, comparisons of the results of aggregate abrasion tests with tests of abrasion resistance of concrete do not generally show a direct correlation. The wear resistance of concrete can be determined more accurately by abrasion tests of the concrete itself. The Los Angeles machine can also be used to test the differences in abrasion resistance between various particles in the aggregate. If the test shows that wear is high at first and then gradually diminishes, it indicates that one element is nonresistant to abrasion.

Cleanliness and Harmful Substances

Most specifications limit the amount of organic impurities, silt and clay particles (finer than No. 200 sieves), coal, lignite, and lightweight and soft particles that are permissible in aggregates. Some tests for cleanliness and harmful substances include:

1. Test for material finer than the No. 200 sieves. In this test, the aggregate is washed vigorously in water, and the wash water containing the fine material in suspension is passed over a No. 200 sieve (sieve openings of 0.0029 in.).
2. Test for friable particles in aggregate. Material that can be pulverized by the fingers is hand pulverized and separated from the sample by sieving.
3. Test for lightweight pieces. Aggregate samples in which poor performance is caused by lightweight pieces are usable when the lightweight particles are removed by beneficiation or are present in small quantities. A limit is frequently placed on the amount of lightweight material when such deposits are used. The sample is placed in a liquid on which the light particles will float and in which suitable particles will sink. These are then separated. The round particles are dried and weighed.
4. Test for organic impurities in sand. Organic materials in sand may interfere with normal cement hydration. Specifications usually require that fine aggregates be free of injurious amounts of organic impurities. To perform the test, the sample is placed in a standard sodium

hydroxide solution and, after 24 hr, the color of the liquid is observed. A dark-colored liquid indicates the presence of impurities.

5. Sand-equivalent test. Sand sample is agitated in a weak calcium chloride solution, and the quantity of colloidal clay is determined, The test provides information on both the amount and the activity of the clay.

Hardness

Hardness can be established by the scratch test for coarse aggregates. It is used to identify materials that are soft and incapable of satisfactory load.

Other Properties

In addition to the scratch test, there are a number of other tests that are not normally specified and are seldom performed. These include tests for toughness, compressive strength, elasticity, particle shape, surface texture, porosity, pore structure, permeability, specific heat, thermal diffusivity, and thermal expansion. Some of these tests may be needed if special aggregate problems arise or if special characteristics are required in the finished concrete.

PRACTICAL SELECTION OF AGGREGATES

High-quality aggregate should contain particles that are free from fractures, abrasion resistant, favorably graded, and not flat or elongated. The surface texture of the particles should be relatively rough with little or no tendency to absorb water. The aggregate should contain no minerals that interfere with cement hydration or that react with cement to produce expansion. Such ideal aggregate is always difficult to locate.

The problem is to appraise locally available, economical aggregates and to decide which ones best meet job requirements. A number of tests have been described for use in specifying certain common requirements and in checking to see these are met. However, tests are not a completely reliable measure of field performance. A better indication of aggregate quality is a good record of satisfactory service. A structure that is completely sound after 10 or more years of service is an endorsement of all the materials, including the aggregate, used in the structure. To make this service record useful, it is necessary to know how and with what ingredients the concrete was made. Records of the cement, aggregate, and water contents of the concrete in the building must be dependable.

It is important to check that the quality of the material one can obtain now is as good as that used in the building with the good experience record. If a structure containing the aggregate being investigated has a poor record of service—one of deterioration—this does not necessarily mean that the aggregate is unsuitable for other uses. The failure could have been due to improper mix proportions, curing, or other causes. Once again, well-maintained records are extremely important.

Where the services of a petrographer are available, the appraisal of an aggregate's service record can be made on a much more scientific basis. Petrographic examination provides information on a wide variety of properties, frequently making it possible to compare samples from a newly developed source with others of known service record.

In actual practice, all of the qualities of an ideal aggregate may not be needed. The cost of specially selected aggregate comes high. Some general guides for avoiding high costs while still selecting an aggregate meeting job requirements include:

1. Excellent concrete can be made with aggregates differing quite widely in grading characteristics.

   It is important that once the grading is established, it be maintained constant within rather close tolerances until the entire job is finished. Otherwise, effective job control is impossible. If after establishing optimum proportions for fine aggregate, coarse aggregate, and cement, the fineness of the sand or the percentage of undersize in the coarse aggregate increases, the water requirement for proper slump will increase. If water is added without compensating adjustments in cement, strength and durability will be reduced. A decrease in the fineness of the sand can adversely affect the workability of the concrete by introducing harshness. Fluctuations in fineness also introduce an added difficulty in achieving uniform air entrainment. Abrasion resistance of aggregate determines the stability of gradation during handling and mixing.

2. An aggregate with unfavorable particle shape should not be rejected in favor of a more expensive aggregate with better particle shape if the cost of additional cement required for the first type of aggregate is less than the extra cost of the second.

3. An aggregate contaminated with organic material to such an extent that the contamination interferes materially with the setting of the cement should not be used.

4. An aggregate that will not produce concrete of the required strength should not be used. If required strength can be achieved only with an

excessively high cement factor, use of the aggregate is probably uneconomical.

5. Material for use in concrete to be exposed to severe freezing and thawing should be proven capable of producing frost-resistant concrete either by service record or freezing-and-thawing tests.

6. Material for use in concrete to be exposed to severe weathering with the added requirement of defect-free appearance should be essentially free of particles that are soft or friable, that have unfavorable capillary absorption, or that produce staining during weathering.

7. Material containing or consisting of substances that might react with alkalies in the cement to cause excessive expansion should not be approved for use in concrete to be exposed to wetting, unless the use of low-alkali cement is also required. A proven pozzolan may sometimes reduce this type of expansion.

8. In unusual circumstances the aggregate user may wish to obtain materials with particular thermal or elastic properties; in such cases, a premium price for the aggregate may be expected.

The entire subject of aggregates for long-lasting, strong concrete can be summarized by five points. They must be (1) chemically as well as physically clean, (2) hard, (3) durable, (4) properly graded, and (5) nonreactive to alkalis.

CHAPTER 7
AIR ENTRAINMENT

INTRODUCTION

During normal mixing action, a small amount of air is always entrapped in a concrete mix. This is derived from air present in aggregates and cement, air dissolved in mixing water, and air brought in by the mechanical action of the mixer. Air from these sources is neither stable nor useful because the bubbles are relatively large, widely separated, and of irregular shape.

In 1932, it was discovered that improved workability and plasticity of concrete resulted from the deliberate introduction of semimicroscopic bubbles of air. Such air lubricates a mix, permits substantial reductions in mixing water, and provides more workable concrete.

Subsequent tests showed this concrete to be extremely durable, particularly when subjected to freezing and thawing in the presence of water and salt solutions. Early experiments also showed that the entrained air almost eliminated segregation and bleeding of the concrete and produced a substantial increase in the workable characteristics of the mix (Fig. 7-1).

Air entraining is defined as the capability of a material or process to develop a system of minute bubbles of air in cement paste, mortar, or concrete. Unlike entrapped air, intentionally entrained air is in the form

Figure 7-1 This pair of microphotographs, taken at the Portland Cement Association's research and development laboratories near Chicago, show typical cross-sections of pavement slab using normal concrete (right) and air-entrained concrete (left). Notice the tiny air bubbles in the air-entrained specimen. Their actual size can be compared with a common straight pin that has been laid over the concrete.

of disconnected bubbles having diameters measured in thousandths of an inch, closely spaced, and well distributed throughout the paste.

Microscopic examinations of hardened concrete indicate that as many as 400 to 600 billion bubbles are entrained in a single cubic yard of concrete with an air content in the 3 to 6% (by volume) range.

Entrained bubbles are both relatively stable and useful for improving the properties of concrete. Stability is imparted to these individual bubbles by the formation of an insoluble layer derived from the reaction of lime and air-entraining agent that envelops the air bubbles during mixing. They also stick to cement particles preventing them from escaping to the surface.

Air can be entrained in concrete by two methods: air-entraining cements or air-entraining agents. Air-entraining materials include salts of wood resins, synthetic detergents, salts of sulfonated lignin, salts of petroleum acids, salts of proteinaceous materials, fatty and resinous acids and their salts, and organic salts of sulfonated hydrocarbons. The material can be interground with cement clinker, thereby producing air-entraining cements, or it may be introduced directly before or during mixing as an agent.

The main advantage of using air-entraining cement rather than adding the air-entraining agent separately is convenience. The addition of another component at the mixer and the problem of accurately measuring small amounts is eliminated when air-entraining cements are used. However, air-entraining cements also have disadvantages. Most of them entrain less air than is required for durability, and frequently an admixture must be used anyway. When air entrainment is excessive, a non-air-entraining cement may be necessary in combination with the air-entraining cement to reduce air content to the proper range.

The main advantage of using air-entraining agents (as opposed to air-entraining cements) is flexibility. Alterations can easily be made to compensate for various conditions that influence the air content or for operations where a variety of concrete mixtures are produced. The engineer or inspector can usually maintain better control of the concrete mixture. Air-entraining admixtures must be dispensed accurately using devices that allow the fluid to flow at a given rate for a set time, that allow a previously measured quantity or weight to flow into the mixer, or that use a flow-meter to measure a given amount.

EFFECTS

Effects of air entrainment on concrete in the plastic state include (1) improved workability, (2) placeability, (3) reduced bleeding, and, in some instances, (4) lower cost. The billions of minute air bubbles act as a lubricant and have an effect on workability similar to adding additional water. In structural work the mix easily assumes intricate shapes. It flows easily around closely spaced reinforcing bars. Since the volume of the paste is increased, lean mixes and mixes with angular or poorly graded aggregates are more workable. Better finishing usually results because greater plasticity allows a smoother, faster finish. Also, finishers can usually start sooner due to reduced surface water. Occasionally troweled finishes may blister or surfaces may crust while drying conditions prevail.

In a sense the bubbles can be considered to be a third aggregate. Because of their small size, they act as fines, cutting down the amount of sand needed. Because air entrainment increases slump, it is possible to decrease the amount of water to get higher strengths without affecting workability. Thus, air-entrained concretes will have lower water-cement ratios than non-air-entrained concretes, and reductions of strength that generally accompany air entrainment are minimized. However, some reductions in strength may be tolerable in view of other benefits. These reductions become significant only in mixes containing more than about

550 lb of cement per cubic yard. In leaner or harsher mixes, strengths are generally increased by entrainment of air in proper amounts.

Attaining high strength with air-entrained concretes may be difficult at times. Both air-entrained and non-air-entrained concretes have increased water demands when slumps are maintained constant as concrete temperatures rise. Even though a reduction in mixing water is associated with air entrainment, mixtures with high cement contents require more mixing water; subsequently, the increase in strength expected from the additional cement is offset somewhat by the additional water added. Also, with some aggregates, it is not possible to secure extremely high strengths with air-entrained concrete. In all such cases, however, other benefits, such as improved durability and workability, are not impaired.

Bleeding in concrete is cut approximately in half by entrained air because the many tiny bubbles greatly increase the surface area of the paste, making it more difficult for water to migrate upward; in addition, subsurface voids act as "relief pressure tubes." This action reduces the adverse effects of a higher water-cement ratio at the surface of slabs and of laitance forming on concrete surfaces; it also permits earlier finishing.

The air-void system in the mortar imparts a buoyant action to the cement and aggregate particles that prevents settlement. Less segregation occurs, and, with finished concrete, more attractive surfaces are possible. Reduced segregation is critical when concrete is pumped with air equipment. The lubricating effect of air-entrainment demands less work on the part of the pumping mechanisms.

In hardened concrete, the outstanding attribute of air entrainment is resistance to weathering and scaling. A properly designed mix with the lowest possible water content will improve the durability of the concrete (Fig. 7-2).

Concrete exposed to repeated cycles of freezing and thawing is sometimes damaged or destroyed. The principal forces causing such damage are internal hydraulic pressures created by the expanding ice-water system and the growth of ice crystals in capillary cavities. When water in a saturated capillary cavity or pore freezes, the expansion produced in the ice-water system requires a dilation of the cavity of about 9% of the volume of the water which freezes. This volume of water is forced out of the cavity and into the surrounding paste, producing hydraulic pressure. The magnitude of the hydraulic pressure depends on:

1. The distance from the cavity to a point of pressure relief.
2. Degree of saturation.

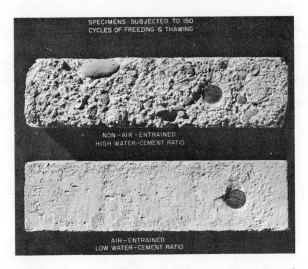

SPECIMENS SUBJECTED TO 150
CYCLES OF FREEZING & THAWING

NON-AIR-ENTRAINED
HIGH WATER-CEMENT RATIO

AIR-ENTRAINED
LOW WATER-CEMENT RATIO

Figure 7-2 Specimens subjected to 150 cycles of freezing and thawing.

3. Rate of ice formation.
4. Permeability of the intervening material.
5. Degree of elastic accommodation of the material around the cavity.

If the pressure developed is too high, the paste will be damaged. Of the five conditions listed above, only the first—the distance from the cavity to a point of pressure relief—can be modified to the extent necessary to prevent damage under severe freezing and thawing conditions. The controlled entrainment of air in concrete serves this purpose.

In any system of voids, moisture tends to move from larger voids to smaller ones because of surface tension. The entrained-air voids, while very small, are far larger than the capillary voids and normally remain essentially free of moisture. They serve as points of pressure relief when moisture is forced out of the capillary voids during freezing. Upon thawing, this moisture is drawn back into the capillary voids by capillary force. Thus, an automatic pressure relief system is built into the concrete.

If enough entrained air voids are present so that no capillary void is more than about 0.008 in. from an entrained-air void, destructive damage due to freezing and thawing will not occur in the paste of normal concrete under normal exposure conditions.

Scaling is caused by severe freezing and thawing conditions and by the wide use of salts as deicing chemicals. Salt scaling can be caused by the direct application of salt to concrete or by the indirect application of salt drippings from automobile undersides. Such damage can be prevented almost entirely if the concrete is air entrained, properly cured, and made with adequate cement content, low water-cement ratio, and sound aggregates.

These conclusions about the durability of air-entrained concrete are the result of thorough, long-term studies by several different sources. In 1941, the Portland Cement Association initiated a study of cement performance in concrete in which construction variables and exposure conditions were controlled to permit direct comparisons between cements being tested. Twenty-seven cements with wide differences in chemical composition and fineness were used in full-size construction, nearly job-size structures, and laboratory-size specimens, for representatives exposures in projects widely scattered throughout the country. All five ASTM C150 types of cement and ASTM C175 air-entraining types IA, IIA, and IIIA were used.

The conclusion reached was that air entrainment is an important requirement of frost-resistant concrete and that adjustments in either the chemical composition of cement, fineness of grinding, or methods of manufacture have relatively little effect on increasing frost resistance of concrete as compared to air entrainment.

The Bureau of Public Roads has sponsored several extensive studies of the durability of concrete with air entrained by different agents. In these studies, slabs, $16 \times 24 \times 4$ in., with a raised edge or dam around the perimeter of the top surface were used as test specimens. These slabs were stored out of doors. During cold weather, when freezing was expected, the top surface of each slab was covered with $1/4$ to $1/2$ in. of water. After the water had frozen, commercial calcium chloride flakes were applied uniformly at the rate of 2.4 lb per square yard of surface.

The slabs were examined periodically and rated for surface scaling. Ratings were based on observations of scaling extent and depth. All of the air-entraining agents used were effective in varying degrees in delaying the start of significant scaling.

Air contents for both air-entrained and non-air-entrained samples were plotted against the number of cycles required to reduce the modulus of elasticity by 50% (Fig. 7-3). All the non-air-entrained concretes had low resistance to freezing and thawing regardless of composition or fineness of the cements, cement content, or water-cement ratio of the concretes. As the air content of the concrete was increased by the use of an air-entraining agent, resistance to freezing and thawing also increased. Con-

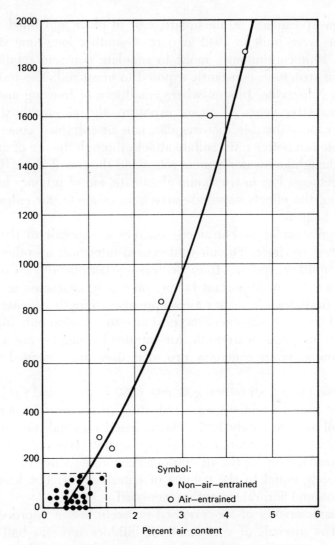

Figure 7-3 Cycle of freezing and thawing for 50% reduction in dynamic modulus of elasticity.

cretes with air contents increasing from 3 to 6% show resistance even after 1250 or more cycles of freezing and thawing. These results demonstrate that the air content of the concrete is far more significant with regard to freeze-thaw resistance than the fineness or composition of the cement.

Box-type specimens, 30 in. square, cast in place and filled with sand and water, were built in 1942 as part of another long-time study. The sand was kept continuously moist to simulate concrete retaining walls and similar structures constantly exposed to moist soil. The test plot was located at Naperville, Illinois, where conditions of freezing and thawing are severe. After years of severe exposure, the air-entrained concrete proved far more durable than regular, non-air-entrained concrete.

Concrete can better resist sulfate attacks through the use of air entrainment. Although low water-cement ratios and the use of Type II, Type V, or other cements low in tricalcium aluminate, are of primary importance in lessening the effects of sulfate attack on concrete, air entrainment is extremely helpful.

Watertightness of air-entrained concrete is superior to that of non-air-entrained concrete. The air-void system interrupts an otherwise continuous capillary system, thus decreasing permeability. Compressive strength is the most important factor controlling concrete's resistance to abrasion. Resistance increases as compressive strength increases. The air content of concrete influences its resistance to abrasion only insofar as it affects the compressive strength. Air-entrained concretes are as resistant to abrasion as plain concretes provided they are designed for equal strength.

For usual class A structural concrete (560 lb per cubic yard and 3-in. slump) air entrainment causes a reduction in compressive and flexural strength of approximately 10%. This is based on conditions of approximately equal cement content and slump for both types of concrete, and with the sand content of the air-entrained concrete reduced by an amount approximately equal to the volume of entrained air. For leaner mixes, compressive and flexural strength is increased.

The characteristics of air-entrained concrete insure improved surface texture. The myriads of very small air bubbles serve as ball bearings; they give better compaction and better consolidation when producing concrete masonry. They enable a harsh, dry-block mix to move into place against form faces. Corners and edges are sharper, web cracks are eliminated, and stripping operations produce a sheen on block faces that greatly improves appearance. Reduced bleeding at horizontal surface joints minimizes the extent of costly surface preparation to remove surface laitance. The same general advantages of air entrainment can be obtained in all concrete products: cribbing, cast stone, precast piles, and so on.

AIR CONTENT

The optimum air content of a concrete is a balance point between increased durability and reduction in strength. This is considered to be the minimum air content; beyond it, further increases in air result only in marginal improvement in resistance to freezing and thawing.

The amount of air to be entrained also depends on the intended use and desired weight of the concrete. Structural concretes should have less air than insulating types.

The desired air content varies with the maximum aggregate size. The nature of the coarse aggregate determines the amount of mortar needed for workability. The larger the aggregate, the less mortar needed. Since entrained air exists in the paste, the less mortar needed the less air required. The mortar component of air-entrained concrete should be adjusted so it contains about 9% air.

If maximum coarse aggregate is:

$1\frac{1}{2}$ to $2\frac{1}{2}$ in., specify . . . . 5% air ±1%

$\frac{3}{4}$ to 1 in., specify . . . . 6% air ±1%

$\frac{3}{8}$ to $\frac{1}{2}$ in., specify . . . . 7$\frac{1}{2}$% air ±1%

It is not feasible to hold the percentage absolutely constant because air content is affected by many factors: maximum size of coarse aggregate, type and gradation of aggregate, hardness of water, length and means of mixing, brand of cement, concrete temperature, etc. A common specification allows for variation of ±1%.

Factors Affecting the Amount of Entrained Air

Several factors affect air content. Mix proportions, aggregate size and gradation, mixing operations, concrete temperature, use of other admixtures, and vibrating and finishing procedures all have their unique effects.

Mixing action is the most important factor in the production of entrained air in concrete. Uniform distribution of entrained-air voids is essential for scale-resistant concrete. Nonuniformity may result from inadequate dispersion of the entrained air during mixing. The amount of entrained air varies with the type and condition of the mixer, the amount of concrete being mixed, and the rate of mixing. Stationary mixers and transit mixers produce different amounts of entrained air due to inherent differences in mixing action and time. An increase in air content may occur if a mixer is loaded to less than rated capacity and a decrease may result from over-loading. More air is entrained as the speed of mixing is increased.

The changes in air content with prolonged agitation can be explained by the relationship between slump and air content. When initial slump is greater than 7 in., air content under continued agitation increases while slump decreases to about 6 to 7 in. Continued agitation will further decrease slumps and air contents as well. For initial slumps lower than about 6 to 7 in., air content decreases as slump decreases with continued agitation.

Aggregate gradation and mix proportions affect both air-entrained and non-air-entrained concretes (Fig. 7-4). Tests illustrated in this graph were made with concretes having slumps of 2 to 3 in. and cement contents of 375, 515, and 655 lb per cubic yard. The quantity of air-entraining agent per pound cement was kept constant. There was little change in air content when the maximum size of aggregate was increased above $1\frac{1}{2}$ in. For aggregate sizes smaller than $1\frac{1}{2}$ in. the air content increased sharply

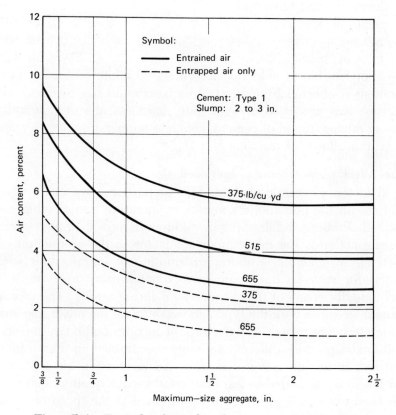

Figure 7-4 Typical relationship between aggregate size, cement content, and air content of concrete.

as the aggregate size decreased. Figure 7-4 also illustrates the effect of cement content on the amount of entrained air. Richer mixes entrain less air while leaner mixes entrain more. This relation applies whether the air is incidentally entrapped or intentionally entrained.

Fine aggregate (sand) content of a mix also affects the percentage of entrained air. Increasing the amount of fines causes more air to be entrained for a given amount of air-entraining cement or agent. Fine aggregate particles in the middle sizes result in more air than the very fine or coarse sizes. Fines passing No. 100 and especially No. 200 sieves decrease air entrainment.

Concrete temperature must be considered. Less air is entrained as the temperature of the concrete increases. The effect of temperature also becomes more pronounced as slump is increased. The effect of temperature is especially important during hot-weather concreting when concrete might be quite warm. Increasing the mixing temperature from 20° to 100°F may reduce the air content 25%, while reducing the temperature to 40°F may increase the air as much as 40%. In cold weather the admixture quantity should be reduced to prevent loss of strength.

Vibration, a method of consolidating fresh concrete, should be shortened and kept reasonably constant since it tends to dispel the air. A normal amount of vibration (about ½ min) reduces initial air content about 10%. When vibrated 1 min, the air content is lowered 15 to 20%. In addition, internal vibration reduces air more than external vibration.

Certain coloring agents (especially carbon black and black iron oxide) and other admixtures add a large percentage of fines to the concrete, rendering the air-entraining agent less effective. Generally, air-entraining agents should be added to the mixer separately from other admixtures unless tests have shown this precaution to be unnecessary.

TESTING FOR AIR CONTENT

Merely specifying air entrainment is not enough in itself. The widespread use of air-entrained concrete in the United States and Canada has led to the development of a number of methods of determining air contents in plastic and hardened concrete.

For plastic concrete, air determinations should be made by competent inspectors at intervals deemed necessary to insure compliance with project specifications. If two consecutive determinations of separate batches at the beginning of a day's run show adequate amounts of air, an occasional check (one per five loads) should provide sufficient evidence of 'continuing compliance. Samples for air tests should be taken directly from the discharge chute of the ready-mix truck.

When air-entraining agents are used, it is necessary to control the air content of the concrete within approximately ±1% limits to maintain uniform workability, strength, and durability. The quantity of air-entraining agent to be used should be adjusted up or down as indicated by the results of the test.

The air content of concrete is the relationship between the volume of air in the paste and the total volume of the concrete. It is normally expressed by the equation:

$$\text{percent total air} = \frac{\text{volume of air in paste} \times 100}{\text{total concrete volume}}$$

Test methods described here can measure only the total percentage of entrapped (large bubbles) and entrained air (small bubbles); one specific kind of "air" cannot be distinguished from the other through testing.

There are three distinct test methods used to determine the air content of plastic concrete. These are the gravimetric, the pressure, and the volumetric methods of air content determination.

Gravimetric Method of Air Content Determination

Later in the text, the absolute volume method of concrete mix will be introduced. One factor in this design procedure is determination of the absolute volumes of the water, cement, sand, and coarse aggregate in a match of concrete. The absolute volumes of one of these materials is that volume which the material occupies in the concrete. For example, if one could mix a match of mortar and determine its volume, then mix coarse aggregate into the mortar and determine the volume of the resulting concrete, the difference between the volumes of mortar and concrete is the absolute volume of the coarse aggregate which was added to the mix.

Normally, absolute volumes of materials are not found by this direct a method. Instead, they are determined from calculations involving the weight of the material and its specific gravity. (Determination and use of specific gravity values will be covered in future chapters.) Specific gravity is a ratio of the weight of a material to the weight of an equal volume of water. For illustrative purposes, only the use of absolute volumes in gravimetric air content determination will be considered here.

Example of Gravimetric Air Content Determination. Assuming that a small concrete operation has a 1-cu yd mixer and consistently mixes concrete with the batch weights given in Table 7-A, and knowing the specific gravity of each material, it is possible to calculate the absolute volume of each concrete component. These values are given in Table 7-A.

With the information in Table 7-A, it is relatively simple to determine the hypothetical unit weight of the concrete if it contains absolutely no air (UWN is a convenient abbreviation). This value is the sum of the weights of all the solid and liquid concrete components divided by their calculated total absolute volume.

$$\text{hypothetical airless unit weight (UWN)} = \frac{3944 \text{ lb}}{25.86 \text{ cu ft}} = 152.5 \text{ lb/cu ft}$$

The actual unit weight of the mixed concrete will be less than this hypothetical value; all concrete contains some air, a material that occupies volume but has no significant weight. The actual unit weight of the concrete (UW) is needed to calculate the concrete air content by gravimetric means. It may be determined by compacting concrete into a container of known volume, weighing the concrete filling the container, and determining the actual unit weight of the concrete by dividing the concrete weight (in pounds) by the volume of the container (in cubic feet).

Assume that the actual unit weight of the concrete in the preceding example was measured and found to be 143.0 lb per cubic feet. The air content of the concrete may then be determined gravimetrically using the hypothetical solid unit weight of the concrete (UWN) and the actual concrete unit weight (UW) by the formula:

$$\% \text{ air} = \frac{\text{UWN} - \text{UW}}{\text{UWN}} \times 100$$

In this example (UWN = 152.5 lb/cu ft; UW = 143 lb/cu ft), the calculated gravimetric air content is:

Table 7–A
CONCRETE BATCH QUANTITIES

	Batch Weight lb	Absolute Volume cu ft
Cement	564	2.82
Water	275	4.41
Sand	1055	6.33
Gravel	2050	12.30
Total	3944	25.86

$$\% \text{ air} = \frac{152.5 - 143.0}{152.5} \times 100 = \frac{9.5}{152.5} \times 100 = 6.2\%$$

The gravimetric air content depends upon a number of factors that cannot be measured accurately. For this reason, air contents determined gravimetrically will be approximate and not as accurate as those found by other methods which measure by direct test.

Air Content Determination by Pressure Tests (ASTM C231)
Air is a gas and obeys certain physical principles. These principles, the "gas laws," relate volume changes to pressure changes in any entrapped gas. Concrete technicians use various kinds of "pressure meters" to directly determine the air content of a concrete sample. These air meters subject the concrete to external pressures, then measure any change in volume. Only air in appreciable volume will change under such small pressure increases. Through suitable calibration and design based on the gas laws, these pressure meters can give direct readouts of concrete air content. The two basic kinds of pressure meters are the "high silhouette" and the "low silhouette."

High-Silhouette Air Meters. Figure 7-5 is a schematic drawing of this type of air meter. It shows (1) the meter bottom or concrete bowl in which the concrete to be tested is compacted and struck off, (2) the air-meter top, which resembles an inverted funnel, and (3) accessories such as the pressure gage, clamps, air pump, and the top cap. As shown in the drawing, the portion of the air meter above the concrete sample is filled with water. This portion necks down into a narrow transparent tube. Small changes in the combined air meter contents of concrete and water result in a significant change of water level in this tube. When the concrete bowl is filled, the air meter assembled, and the entire device filled with water, one substance in the meter is very compressible—the air that is either entrapped or entrained in the concrete. The water and cement are incompressible, and the fine and coarse aggregate have little compressibility at this operating pressure. The quantity of total air entrained and trapped in the concrete is determined through a simple process. Air is pumped into the meter until the pressure gage at the very top registers a fixed predetermined "operating air pressure." The percent for this pressure is indicated by the air-content marks on the tube level. If the concrete in the meter contains a great deal of air, the air-meter charge will be compressed considerably and the surface of water in the stack

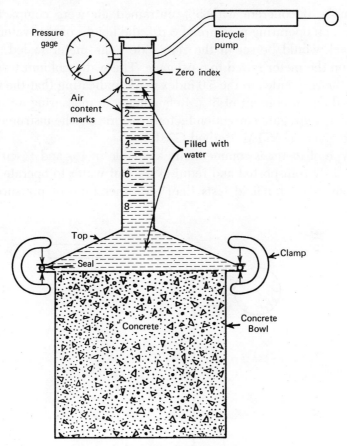

Figure 7-5 High-silhouette type ari meter, concrete air test setup.

will drop. If the concrete contains little air, the air meter charge will be mostly incompressible and the surface of water in the top stack will remain approximately constant. The basis of this test is the principle of physics known as Boyle's law ($P_1V_1 = P_2V_2$), which governs the relationship between the pressure and volume of a gas.

A rulerlike scale is set close to the surface of the transparent stack with the zero index at the top and numbered divisions increasing downward. The air meter, open and unpressurized, is initially filled so that the water level in the stack is set at the zero index mark. Increasing this initial pressure to the operating air pressure causes the water level in the stack to fall to an index mark equivalent to the percent air in the concrete.

For example, if concrete with 5% entrained air were compacted in a meter with an operating pressure of 8 psi, the height of the water column in the stack would be set at the zero index, the meter sealed, and the pressure on the meter raised to 8 psi gage. The water column would then fall from the zero index to the 5.0 index mark, indicating that the concrete charge had an air content of 5%. Refinements in measuring air contents, determining aggregate correction factors, calibrating the instruments, and so on, are given in ASTM Method C231.

This type of meter is commercially available in ½- and ¼-cu ft sizes. It is not easily transported and requires a lot of water to operate, making it less convenient for field tests than the other type of pressure meter.

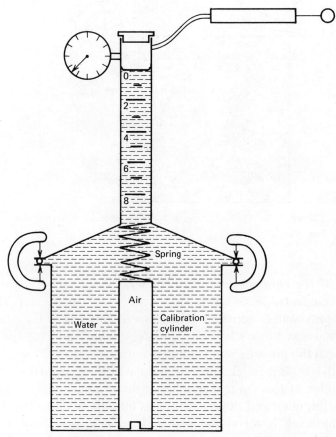

Figure 7–6 High-silhouette type air meter, calibration setup.
setup.

However, its simplicity, ruggedness, and accuracy make it extremely desirable for concrete work in laboratory situations.

The calibration of this type air meter involves determining the one operating pressure that will cause the water to decrease from the zero index to a stack reading that absolutely matches the air content of the concrete compacted in the air-meter bowl. Figure 7-6 shows the setup normally used for the calibration of this air meter. A small cuplike cylinder is inverted in the empty air meter and held in that position by a spring action against the top of the meter. When the air meter is filled with water, a full cup of air is trapped inside the cylinder. The "percent air in the cup" may be determined as a percent of the volume of the bowl (normally filled with concrete during air tests) using the following relationship:

$$\% \text{ air in cup} = \frac{\text{volume of cup}}{\text{volume of bowl}} \times 100$$

To obtain a reasonably good air meter calibration and find the required operating pressure of the air meter, it is only necessary to determine, by actual trial, the gage pressure required to compress the air in the cylinder enough that the reading on the stack, in percent, matches the "percent air in the cup." For very critical test work, refinements in this calibration procedure are required. These are given in ASTM Method C231.

Because this meter is a pressure device, an aggregate correction factor must be made; the pores inside most aggregates are not completely saturated with water but are filled with a mixture of both water and air. Air in aggregate pores is compressed during the operation of the air meter. Compression effects of aggregate pore air must be separated from that of intentionally entrained air in the concrete because entrained air, by definition, occurs as voids in the paste. Aggregate compressibility does not fall into this classification. The use of an aggregate correction factor permits calculation of the correct air content in concrete.

To determine the aggregate correction factor, fine and coarse aggregate of the same type, gradation, and moisture content as those to be used in the concrete are placed in the air meter, covered with water, and very thoroughly stirred to remove any air bubbles that may stick to the external surfaces of aggregate particles. An air test of the aggregate is then made in the same manner as for air content of concrete. The "percent air" determined on the aggregate only is equivalent to the compression effects on pore air in concrete aggregate during air tests. The aggregate correction is made by deducting the "percent air" determined on the aggregate only. The difference is the "net air" content of the concrete and is the actual air content within the paste portion. Further instructions for aggregate correction factor determination are given in ASTM Method C231.

Low-Silhouette Air Meters. The high-silhouette air meters are designed to measure air content as a change in the volume of the air meter contents in response to a change in pressure inside the meter. The "low silhouette" air meter measures air content of a concrete sample as a function of the change in pressure that occurs when a small volume of compressed air is jetted into the air meter. The Press-ur-Meter, the Techkote meter, and a number of other similar devices are low-silhouette meters used in the field and laboratory.

As indicated in Figure 7-7, the main components of the low-silhouette meter are a bottom container, in which the concrete sample is compacted, and the top portion of the meter, which contains the air-sensing devices.

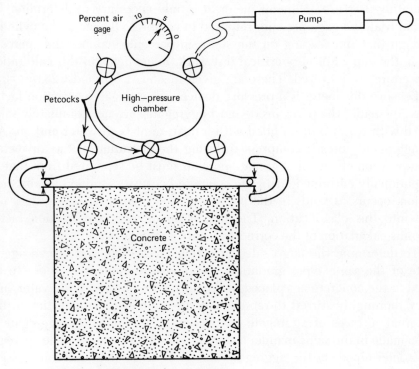

Figure 7–7   Low-silhouette type air meter.

The air-meter top consists of the meter closures: clamps and seals, a fixed-volume pressure chamber, an air pump, an air pressure gage with a special dial, and various valves to connect the top portion of the air meter to the concrete bowl and to the outside.

The basic air-sensing method that these meters use determines the amount of pressure drop when air pumped to a high pressure in the top chamber is released into the bottom concrete-filled portion of the air meter. If the pressure loss is great, the air content of the concrete is relatively high; if the pressure loss is slight, the concrete contains little air. These reactions are subject to Boyle's law, the principle used in designing the pressure gage so that it can directly register, in percent, concrete air that is compacted in the air-meter bowl.

The low-silhouette air meter's portability makes it an ideal instrument for field use. Its overall bulk is smaller than high-silhouette meters, it requires less water and concrete for testing, and it is easily calibrated on the job. This air meter is the type most commonly used in the field.

In conducting an air meter test with this device, the following procedure is used.

1. The concrete bowl is filled with concrete, and the surface is struck off level with the top surface of the bowl.
2. The meter is assembled with all petcocks open except the one connecting the pressure chamber and the concrete bowl.
3. The space at the top of the concrete bowl is filled with water through a petcock that connects the bowl and the atmosphere. Since the air in this space is not purged during the water-filling process, it will count as entrained air; this operation is critical and should be done thoroughly.
4. The petcocks to the pressure chamber are closed and the pressure is pumped to the required starting point. The water-purging stopcocks to the concrete bowl are closed.
5. Quickly thereafter, the petcock between the pressure chamber and the concrete bowl is opened. The air-content reading on the pressure dial is noted as soon as it comes to rest. This is the "gross air content" of the mix.
6. The aggregate-correction factor (the previously mentioned effect of pore-air compression) is deducted from the gross-air content (the air in the paste plus the compressibility effects of air in the aggregate pores). The resultant figure is the net air content of the concrete, the specification value for air content. As with the high-silhouette type meters, the aggregate correction factor is found by determining the air

content of the meter full of aggregate of the same gradation and initial moisture content as aggregate in concrete. Particularly critical to the operation is an extremely thorough stirring and agitation of the aggregate so that no air external to the aggregate particles remains in the concrete bowl.

Calibration of low-silhouette-type air meters should be done according to the operating instructions furnished with the meter. Most of the devices are calibrated by being completely filled (that is, all available space but the pressure chamber filled) with water. A measured amount of water is then emptied from the meter through one of the petcocks connecting the concrete bowl with the outside of the meter. This withdrawal of water is accompanied by the introduction of an equal volume of air within the meter. This volume of withdrawn water and introduced air can be measured in the small volumetric container furnished with the meter, or might be calculated in milliliters or fluid ounces as about 5% of the concrete-bowl volume. Once the water is withdrawn, an air test can be run on the meter's water-air charge, and the percent air read with the meter can be determined. If the tested air content value is above or below the actual percentage of air in the meter, the meter can be recalibrated by altering the zero index point of the meter.

Volumetric Air Meters
One of the old-style procedures for determining concrete-air content is a direct determination through the use of a yield bucket, a hook gage, the rod normally used for concrete test compaction, and a milliliter graduate. The half-cubic-foot yield bucket is filled with compacted concrete to an index mark three-eighths of the way up the inside wall of the bucket. Water is then poured onto the concrete until the bucket is filled to three-fourths of its height. This water level is critical and is controlled by the use of a hook gage. The hook gage is removed, and the water and concrete is stirred into a slurry. Loss of air may be determined by replacing the hook gage and pouring in water from a graduate until the precise initial height of fluid in the unit weight container is restored. Air content in percent is calculated as the volume of concrete (in milliliters) divided by the amount of water added from the graduate (in milliliters), the ratio time 100. This old-style procedure illustrates the technique of "volumetric" air determination in which concrete is reduced to slurry, air is removed, and loss of air is determined volumetrically and then converted to a percent of the total concrete volume.

A newer volumetric procedure is more widely used; it is outlined in

ASTM Method C173. One commercially produced meter meeting this specification is the "Roll-a-Meter." A concrete sample tested with such a meter is quite small (less than $\frac{1}{10}$ cu ft), and the test procedure takes longer and is somewhat more involved than testing with a pressure meter. However, the volumetric method is superior to the pressure procedure for testing lightweight aggregate concrete. Large aggregate correction factors and the complex, time-consuming pressurization of air voids in lightweight aggregate makes testing lightweight aggregate concrete with a pressure meter quite difficult and somewhat inaccurate, even when done by experts. Because volumetric test procedures give the true air content of paste only, no aggregate correction factor is needed with this technique. As a result, the Roll-a-Meter is an extremely popular and effective device for lightweight aggregate concrete testing.

The Roll-a-Meter (Fig. 7-8) consists of a concrete bowl at the bottom, a top equipped with permanently mounted clamps, a screw cap, and a small alcohol-measured jigger with a volume equal to 1% that of the

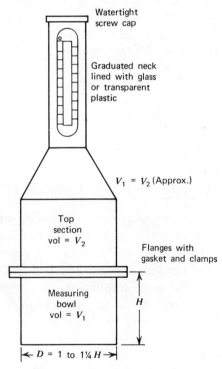

Figure 7–8 ASTM C173 volumetric air meter.

bowl. To run an air test with the Roll-a-Meter, the following test sequence is used:

1. Concrete is compacted in the bowl and struck off flush with the top.
2. The top is attached to the bowl and filled with water to the index mark at the very top of the narrow tube extending it. The cap is then screwed on.
3. The concrete-water-meter charge is reduced to a slurry, and the air bubbles are washed to the top of the slurry. This is done by inverting and reinverting the meter several times, grasping the top tube, and rolling the whole meter back and forth on a flat surface for at least 2 or 3 min. The meter is then brought upright gradually so that all the air meter foam rises into the stack, which has a transparent sidewall calibrated in percent of the volume of the concrete bowl.
4. A good reading is impossible as long as 2 or 3 in. of foam, which contains an appreciable volume of water and cement, sits atop the slurry in the air meter. The foam is broken with alcohol that dissipates the bubbles and sends solids to the bottom of the meter. A good reading can then be taken. Two or three jiggers of alcohol (rubbing isopropyl is best, wood alcohol will do) may have to be added to break the foam. Each jigger of alcohol raises the liquid level in the stack exactly 1%, so a count of the number of jiggers that are added to break the foam should be noted. This number added to the reading on the air-meter determines the percent air in the concrete.

Advantages of the Roll-a-Meter include the fact that it need not be calibrated if it is in usable condition and the fact that there are no moving parts used in determining the air content of the concrete.

TESTING FOR AIR IN HARDENED CONCRETE

The best means presently available for determining air content of hardened concrete are direct optical measurements of air voids that are intersected by the plane of a finely ground section of concrete. This method provides information on total air content and size and distribution of air voids. It has provided the most significant data regarding the factors governing the effectiveness of air in protecting concrete from frost damage.

Linear-Traverse Method

The linear-traverse method involves measuring the chord lengths of air voids that are intersected by a line of traverse on a random plane sectioned through the concrete. The summation of the chord lengths across air voids represents the fractional proportion of air along the total length of traverse and is directly equivalent to the proportional volume of air void volume in the concrete.

Modified Point-Count Method

The modified point-count method is similar to the linear-traverse technique but varies by substituting a series of closely spaced points on a line of traverse rather than performing a continuous integration along the line. The number of points that fall on air voids divided by the total number of points is directly equivalent to the volumetric proportion of air in the concrete.

CHAPTER 8
ADMIXTURES

INTRODUCTION

Admixtures for concrete are not new to the construction industry. In the first century B.C., Marcus Vitruvius, a noted Roman architect and engineer, suggested using blood, hog's lard, or curdled milk in stucco mixtures. Unknowingly he had stumbled on the first air-entraining admixture. Since this earliest discovery, countless other admixtures for concrete have been introduced, each intended to impart some special property—strength, durability, and economy—to concrete.

The use of admixtures has its advocates as well as its critics. Some admixtures have been approved for specific uses by such professional organizations as the American Concrete Institute, the American Society for Testing and Materials, the Bureau of Public Roads, the American Association of State Highway Officials, and the Army Corps of Engineers. Others are reported to be multipurpose materials. Because they are intended to improve concrete in several ways and offer a variety of desired results, they are difficult to categorize. For this reason, principal admixtures and their intended results often tend to overlap.

CONCRETE QUALITY AND ADMIXTURES

Engineers, architects, and contractors working with concrete have felt for years that the best admixture was additional cement. That situation has changed. Instead of a singular solution, a multitude of materials are now promoted and sold for inclusion into concrete. One fact, however, remains prominent: admixtures are not a substitute for quality concrete. They should only be used to modify existing properties to specific job situations.

The growing use and new applications of portland cement concrete, mortar, and grout occasionally require certain properties not anticipated in the past. Even when special properties are not required, a producer may realize that his concrete is lacking in durability, finishability, or workability. Admixtures may provide a solution.

An admixture's effectiveness depends on many factors: type and amount of cement, slump of concrete, mixing time, water and air contents, and batch temperature. Effectiveness varies with mix design.

Quality concrete should be workable, strong, durable, watertight, and wear resistant. While it is not always necessary to have all these properties in any one mix, the more there are present, the more universally acceptable is the product.

Before any admixture is recommended for a concrete mix, its effect on all the properties of the concrete, not strength alone, should be considered. Strength, as vital an asset as it is, may not be sufficient reason for using an admixture; early and ultimate strength should be considered as should possible effects on volume change, wear resistance, and the other general properties of sound concrete.

An admixture should be used only when good and specific reasons warrant its use. Even when it is decided that an admixture is justified or necessary, two precautions should be taken: the admixture must meet the requirements of applicable ASTM or other specifications, and it should be tested in advance with actual job materials and under actual job conditions to predetermine its eventual performance.

TYPES OF ADMIXTURES

Committee 212 of the American Concrete Institute, in its 1971 report, grouped admixtures into five different major categories. Previously, the Committee had classified admixtures in 11 categories. They said, "it is recognized that the classification of admixtures is difficult due to the fact that there are many varieties and types of admixtures and also that some admixtures produce more than one effect."

The difficulty in categorizing admixtures underscores the problem facing a specification writer. If materials are so divergent that they cannot be grouped into standard categories, concrete containing these materials will obviously vary widely in properties. The function of admixtures is to provide concrete special properties that cannot be economically secured by other means; across-the-board use of an admixture, with the exception of an air-entraining agent, could well result in concrete with properties not desired for a specific project.

The most authoritative source of information on the subject is the report prepared by Committee 212 of the American Concrete Institute.[*]

SPECIFICATIONS AND TESTS FOR ADMIXTURES

Before the development of comprehensive specifications and test methods for admixtures, the selection of an admixture was based mainly on limited field experience. If the selected admixture proved satisfactory, it was thereafter designated in the specification by its proprietary name; this is still true of many specifications in use today. If admixtures are required they should be designated according to appropriate ASTM or other specification.

ASTM C260, the standard specification for air-entraining admixtures, for example, stipulates that concrete made with air-entraining admixtures must meet certain requirements for bleeding, compressive strength, flexural strength, resistance to freezing and thawing, time of set, and length change of concrete. The test methods for each of these properties are covered by ASTM C233, the standard method of testing air-entraining admixtures for concrete. Essentially a performance specification not concerned with the source, nature, or chemical composition of the admixture, it is intended to ensure that an air-entraining admixture will do specifically what is claimed for it and, of equal importance, that the other properties of concrete will not be adversely affected.

ASTM C494, the standard chemical admixture specification for concrete, covers water reducers, retarders, accelerators and combinations of water-reducing/retarding admixtures and water-reducing/accelerating

---

[*] "Guide for use of Admixtures in Concrete," as reported by ACI Committee 212, is reprinted in the Appendix in its entirety with the kind permission of the American Concrete Institute. Periodically these reports are revised and updated. While this report is current at this time (1972) it is advisable to check with the Institute for any alterations.

admixtures. The specifications and test methods are patterned after those of ASTM C260. A limit is placed on the specific effect claimed for a given type of admixture, and, at the same time, a requirement that the admixture not adversely affect other properties of the concrete (water requirement, time of set, compressive and flexural strengths, bond strength, length change and resistance to freezing and thawing) is specified. Table 8-A sets forth the physical requirements for chemical admixtures. The specification points out that tests should, if possible, be made using the cement, aggregates, admixture, and proportions that are to be employed on the job, and that time of set tests should be made with the materials at a temperature $73° \pm 3°F$ ($23.0° \pm 1.7°C$).

Another specification in use today is ASTM C618, a tentative specification for fly ash and raw or calcined natural pozzolans for use in portland cement concrete. It is similar in intent to ASTM C260, but contains chemical as well as physical requirements.

These specifications, covering the most common types of admixtures, represent a great advance in providing unbiased evaluation of commercial products. They ensure that an admixture will perform according to claim without certain harmful effects; they effectively show up useless, inferior or harmful products; they provide equal opportunity for all products regardless of nature, source, and chemical composition; and they greatly simplify the problems of the user. One disadvantage is the cost of carrying out the complete series of tests.

To date, ASTM has not developed corresponding specifications for certain other admixture types, such as waterproofing and damp-proofing agents, hardeners, pigments, expanding agents, and wetting agents. Some of these are of questionable value, and a major difficulty is the development of tests that will yield significant information.

GENERAL CONSIDERATIONS

The foregoing general specifications provide an acceptable means for eliminating unsatisfactory admixtures and for developing performance records for potential users or producers. Some problems involved with the availability of materials, achievement of acceptable test records, costs of testing, and the provision of adequate plant and job facilities and controls still exist. The main problem is the actual testing. The specifications are given, but the user or producer must still do the testing. Although some large operators have competent test facilities and sufficient resources to cover the expenses involved, many of the smaller businesses do not.

Table 8–A
CHEMICAL ADMIXTURES FOR CONCRETE (ASTM C494) PHYSICAL REQUIREMENTS

	Type A, Water Reducing	Type B, Retarding	Type C, Accelerating	Type D, Water Reducing and Retarding	Type E, Water Reducing and Accelerating
Water content, max. % of control	95			95	95
Time of setting, deviation from control in hours					
Initial { max.	±1	+3	−3	+3	−3
{ min.	—	+1	−1	+1	−1
Final { max.	±1	+3	—	+3	—
{ min.	—	—	−1	—	−1
Comp. strength, min., % of control					
3 days	110	90	125	110	125
7 days	110	90	100	110	110
28 days	110	90	100	110	110
6 months	100	90	90	100	100
1 year	100	90	90	100	100
Flexural strength, min., % of control					
3 days	100	90	110	100	110
7 days	100	90	100	100	100
28 days	100	90	90	100	100
Bond strength, % of control					
28 days	100	90	100	100	100
Volume change, expressed as % change in length, max., increase over control					
28 days, 6 months, 1 year	0.010	0.010	0.010	0.010	0.010
Durability factor, min., % of control	80	80	80	80	80
Bleeding, increase over control, max. % net mixing water	5	5	5	5	5

Fortunately, most admixture manufacturers are willing to provide technical aid and dispensing facilities in cooperative evaluation programs.

Adapting an admixture to specific materials in a plant should not be limited to laboratory tests. It should include field trials and job-situation tests as well. One of the most important requirements is uniformity of properties of both plastic and hardened concrete. Other specific effects not normally covered by specifications should be noted—finishing properties and effects of variations in weather, for example.

Some admixtures are supplied in solid form, others as liquids. Powders, flakes, and pastes are obviously difficult to distribute homogeneously in a mix. It is preferable to premix the water-soluble type, solid or liquid, with the mixing water. Some, even in the liquid state, form colloidal suspensions with water. They appear to be stable but may coagulate or settle in a tank. In such cases the suspension should be continuously stirred.

Admixture dispensers in a plant must be calibrated frequently. Overdosages can lead to serious field problems. This is particularly critical with retarders where delay in hardening has been known to extend to more than a week and with air-entraining admixtures where excessive air contents reduce strength.

Another problem, often overlooked, is the effect of one admixture upon another under certain conditions (e.g., air-entraining agent plus retarder). If these are introduced into the mix at the same time, or are premixed together in the mixing water, they sometimes react or coprecipitate producing unusual and varied effects. This can be avoided by introducing the two admixtures at different times in the sequence of batching and mixing. Some of the water-reducing admixtures, notably the lignin derivatives, greatly increase the effectiveness of an air-entraining agent.

In the case of air entrainment, a satisfactory amount of air, as determined by an air meter, does not necessarily mean that good durability will naturally follow. Certain natural sands and some lignosulphonate admixtures entrain considerable amounts of air.

A concrete producer's initial delight in finding out how small a dosage of air-entraining agent provides sufficient specified air may be short-lived when the agent's harmful effect on durability becomes known. The true test of the air entrainment should be made microscopically on the hardened concrete; the size and spacing of air bubbles could be more accurately determined in this manner. If the time element permits, actual freezing and thawing tests should also be carried out.

The user of an admixture should set up a rigid specification for the product itself, just as he normally does for any other ingredient of concrete. Each shipment should be tested for uniformity. The admixture supplier should be required to give notice of modifications or formula changes. Some admixtures are claimed to possess built-in safety factors that permit considerable over-dosage with no ill effects. This should not be taken for granted.

The concrete producer should further inform his customer what admixture is present in the concrete and how it will perform. It is traditional that the manufacturer of a finished product can, if he wishes, keep his formula secret; however, plastic concrete is not a finished product, and subsequent operations may involve difficulties if this information is not available.

Water-reducing admixtures are sometimes used to permit a reduction in mixing water used, and, consequently, in cement content, to achieve a given strength. Fly ash or pozzolans are occasionally used in mass concretes, in some cases in conjunction with water-reducing agents. Large jobs with precision-controlled batching on the job can effect substantial savings in this way where responsibility for quality control is not divided. The general use of such admixtures in some ready-mix operations could be subject to additional problems and risks.

Retarders of the lignosulfonate and hydroxylated carboxylic acid types usually reduce water requirements and increase strengths for a given cement content. The difficulty with these materials is in the variability in their effects with variations in mix properties, materials, and weather. In leaner mixes, it is particularly important to have enough cement to produce a dense, impermeable mass. It is essential that when air is added, the mix be redesigned to compensate for this added air and to maintain a constant cement content.

Air-entraining agents generally require a mix proportion using less sand, and therefore economy may be affected. Admixtures that improve workability may facilitate placement of concrete thus reducing an important cost item. The improvement in workability may be attributable to any of the followinig qualities of admixtures: entrainment of air, decrease in surface tension of the mixing water, and particle shape of pozzolanic materials.

Using an admixture primarily to save cement is not good practice. The producer of concrete can probably achieve much greater economy by simply improving the quality of his materials and the efficiency of his operations. Refinements in quarrying, crushing, sizing, stockpiling and handling of aggregates, in mix proportioning, and in quality control and uniformity result in economical production.

SUMMARY—QUESTIONS FOR CONSIDERATION

Before specifying an admixture, the following questions should be resolved:

1. What special property or properties are needed? What alternate methods are there for achieving these properties?
2. Is it necessary to use an admixture? If so, what kind of admixture should be used? Do you know of any possible side effects of this admixture?
3. Is the admixture you are considering covered by ASTM or federal specifications? Do the applicable specifications allow any side effects that could not be tolerated?
4. What evidence do you have that the admixture you are considering performed satisfactorily under similar conditions?
5. Who is responsible for the performance of the admixture, the ready mix producer? the specifier? the admixture supplier? Who must stand the cost of additional finishing time if excessive retardation occurs, the contractor? the specifier? the admixture supplier? If the admixture proves chemically incompatible with other materials in the mix, who must pay for replacement of the concrete?
6. What brand of admixture will you use? How much will you add to the mix?
7. Will you specify a minimum cement content?
8. Based on-the-job observations: Does each batch perform as well as the preceding one? What are the finishing characteristics of the mix? What is the effect of varying the temperature of the concrete materials?
9. How does the addition of the admixutre affect the design of the structure you are building? Does it affect the strength, modulus of elasticity, shrinkage, creep, or fire rating?
10. How does use of the admixture affect the total cost of the concrete? Consider the following: cost of admixture, cost of batching or dispensing equipment and maintenance, admixture handling cost, cost of additional field inspections, resulting time or labor savings.

CONCLUSIONS

In considering admixtures and their effect on the properties of concrete, it is important to remember the following:

1. All admixtures should conform to applicable American Society for Testing and Materials (ASTM) or federal specifications.
2. No one admixture or proportion of such admixture will produce optimum quality in all concrete mixtures under all conditions.
3. An admixture should not be considered a substitute for portland cement.
4. Trial mixes should be made at job temperatures and with job materials to obtain a comparison between a properly designed normal concrete mix and one containing an admixture. In this way, the compatibility of the admixture with other materials and the effects of the admixture on the properties and economy of the fresh and hardened concrete can be observed.
5. If admixtures are used, the dosage should be carefully determined by field tests.
6. A well-designed concrete mix containing an adequate cement content will generally produce concrete with the desired properties. If this is not possible with local materials, a comparison should be made between the cost of changing the basic mix design and the cost of using an admixture. Admixtures could add to the cost and increase the possibility of wide variation in the properties of the concrete.
7. In general, whenever the water-cement ratio of concrete is increased, the effect is detrimental. If an admixture requires additional mixing water, any beneficial effect it may have will more than likely be offset by a decrease in durability, strength, or other property.
8. There is evidence that many chemical admixtures react with cement to form compounds not otherwise present. This is a probable reason why some of them promote shrinkage while also reducing the unit water content.
9. Selecting a specified minimum strength-minimum cement content as the basis for a concrete mix will do more than anything else to assure a quality concrete.
10. In almost all areas of the country, it is economically possible to produce a quality concrete mixture by specifying (a) minimum cement content, (b) maximum allowable water content, (c) size, quality, and gradation of the aggregates, and (d) slump desired at the point of delivery.

SECTION THREE
QUALITY CONCRETE– GENERAL PROPERTIES

CHAPTER 9
WHAT IS CONCRETE?

INTRODUCTION

What is concrete? It is a composite material consisting of a binding medium or glue (cement and water) in which particles of a relatively inert filter material (sand and gravel aggregates) are embedded. It begins as a plastic mixture and gradually hardens into a stonelike mass. The properties of hardened concrete are of greatest concern because they determine the ultimate usefulness of the product. Generally speaking, the material must be strong, durable, watertight, attractive, and economical.

All of these qualities are possible if certain standards are adhered to: good planning and design, good proportioning of quality ingredients in the mix, careful handling and placing, correct finishing, and proper curing. The basic ingredients for both good and bad concrete are the same: cement, water, and aggregate. The quality of the final product is totally dependent on the standards outlined above.

MAKING GOOD CONCRETE

Bad concrete cannot always be immediately identified. Sometimes it takes years for the bad effects in the form of cracking, breakage, deterioration of the surface, staining, or discoloration to become evident. While many of these defects are the result of poor design or choosing the wrong type of concrete for a particular application, most are caused by poor constructional practice.

Assuming that cement, water, and aggregate meet high standards (the water is free from impurities, the aggregate is durable, hard, and clean, and the cement meets or exceeds ASTM standards), the quality of the concrete then depends entirely on what is done with these ingredients. First, the correct amount of each ingredient must be determined. This varies depending on the particular use for which the concrete is designed. Then, the concrete must be properly mixed and transported to the job site. Finally, correct placing and finishing, followed by proper curing, will assure the desired results.

RESPONSIBILITY

The responsibility for good quality concrete rests with three people: the architect or engineer, the supplier, and the contractor. The architect or engineer is responsible for designing the structure, taking advantage of concrete's special capabilities and avoiding its limitations. He must specify the proper concrete for the particular application. The responsibility for quality is the supplier's; he must obtain the proper materials, calculate the right proportions, and mix them adequately. When the material has been transported to the job site, the contractor assumes reseponsibility. He and his personnel must handle the plastic concrete without damaging it, place it correctly, and see that it is properly cured. If any one of the three individuals fails to perform any aspect of his job carefully, the resulting concrete is likely to be of poor quality.

HYDRATION

The study of what happens to cement when it combines with water is very interesting and important. It must change from an extremely fine powder—so fine that several thousand grains can fit on the head of a pin or pass a sieve with 40,000 openings per square inch—to a plastic mixture called paste and, eventually, to a hardened mass. The chemical reaction resulting from the mixture of cement and water is called *hydration*.

For cement to hydrate requires a certain amount of water—theoretically, at least 0.4 g per gram of cement. In order for cement hydration to continue, this water must not be allowed to evaporate. For this reason specifications require some type of curing that supplies additional water or inhibits initial water loss.

Cement and water combined make hydration products. Space is necessary to house these products, and since the space used is that originally occupied by water, it appears that the more water that the mix contained, the more complete the hydration process would be. However, all of the cement will not hydrate even after several years in a saturated atmosphere. Strength can be attained even if all of the cement does not hydrate, and higher strength is actually attained at lower water contents.

When cement and water combine, each particle of cement is surrounded by water. Hydration begins, producing a number of complex chemical substances. The dominant product consists of tiny particles suspended in liquid. This material is known as *tobermorite gel*. Although other products are formed during hydration (crystalline calcium hydroxide, for example), all these products are often lumped together and called "cement gel." This term is apt because the gel is definitely the predominant product.

As hydration proceeds, these gel particles continue to grow until those around one particle of cement become intertwined with those of other particles. At this point, the paste becomes somewhat rigid; initial set has occurred. As hydration continues, the gel particles become more firmly intertwined, and the cement gains stability. Other products of hydration begin to fill the spaces between the gel particles. If the spaces are not completely filled, as is the case when excess water is added to the mix, air spaces or capillary pores remain in the concrete. These result in concrete that is lighter, more porous, less strong, and subject to absorption of water. If the reaction products completely fill all the spaces between gel particles, the cement paste becomes strong and impermeable to water. In comparisons of specimens containing equal amounts of cement but varying amounts of water, the concretes with more water have greater volume but less strength.

The initial period of cement hydration has four stages. (1) Immediately on contact of the cement and water there is a period of about 5 min when chemical reactions occur rapidly. These cause the temperature of the mix to rise rapidly. (2) The rate of heat generation then drops and remains low for about an hour. This is called the dormant period. (3) An increase in heat generation begins, peaking about the sixth hour after mixing. (4) The fourth stage begins as the heat generation starts to drop to a very low rate, usually within 24 hr, and to still lower rates after that. Among cements of various compositions, this pattern varies. The reaction pattern is very sensitive to the proportion of gypsum in the cement. Gypsum acts as a retarder, slowing down hydration for reasonable placement time.

WATER-CEMENT RATIO

Water is essential for hydration. As long as enough of it is present in the concrete, and some cement remains, hydration continues. However, when all water has been combined in the hydration process or the internal humidity drops below about 80%, hydration stops. In such cases, cement particles are not hydrated as completely as possible and the concrete cannot develop its full strength. For this reason curing is essential during the hardening process.

Concrete, like glue, coffee, or soup, requires the right amount of water —too little makes it too thick and too much makes it lack character. Cement requires a water-cement ratio of about 0.4 by weight to complete the chemical reaction that changes paste into a hardened mass. However, a concrete thus proportioned would be so stiff that it would be extremely difficult to place and finish. Subsequently, additional water must be added to make the mix workable. This additional water is called "water of convenience." A water-cement ratio of 0.5 to 0.7 is usually required to make workable paste.

If too much water is added, the mix becomes so fluid that it can be poured into place. However, the time saved in placing will be lost in finishing, and extra water reduces concrete's strength, durability, water-tightness, and resistance to freezing and thawing. Excess water (that which does not participate in the hydration process) evaporates and leaves voids in the concrete.

The correct water-cement ratio depends on the intended use of the concrete. For instance, thin concrete for railings, curbs, ledges, or architectural purposes should have a relatively low water-cement ratio. General purpose concrete can be made with a higher water-cement ratio as can concrete for use in structures of great mass. If the hardened concrete is to be in contact with water, especially sea water, it must be more impermeable; therefore, the water-cement ratio must be lower. Concrete subject to freezing and thawing must be made stronger and more durable than concrete used in more temperate climates. Table 9-A shows water-cement ratios recommended for different applications.

When measuring equipment is designed to measure gallons, it is sim-

Table 9–A
ACI RECOMMENDED MAXIMUM PERMISSIBLE
WATER-CEMENT RATIOS FOR CONCRETE IN SEVERE EXPOSURES[a]

Type of Structure	Structure Wet Continuously or Frequently and Exposed to Freezing and Thawing[b]	Structure Exposed to Sea Water or Sulfates
Thin sections (railings, curbs, sills, ledges, ornamented works) and sections with less than 1 in. cover over steel	0.45	0.40
All other structures	0.50	0.45

[a] Adapted from Recommended Practice for Selecting Proportions for Normal Weight Concrete (ACI 211.1-70).
[b] Concrete should also be air entrained.
If sulfate-resisting cement (Type II or Type V of ASTM C150) is used, permissible water-cement ratio may be increased by 0.05.

ple to convert the reading to weight by multiplying the reading by the weight of 1 gal of water, 8.33 lb (or, for metric measure, 3.78 kg).

The yard is used to express the volume of concrete. A yard in concrete language means a cubic yard; that is, a volume equal to that contained by a cube 3 ft on each side. A cubic yard therefore contains 27 cu ft of concrete.

Cement is the most costly ingredient in concrete unless special aggregates are used. For this reason the quantity of cement should be kept as small as possible without sacrificing the quality of the finished product. The amount of cement needed in any batch of concrete depends on the water-cement ratio of the paste, the consistency or stiffness required in the fresh concrete, the maximum size of aggregate used, the distribution or gradation of sizes of aggregates, the shape and texture of the aggregate particles, and the amount of entrained air in the mix.

AGGREGATE

Aggregate is used as a filler to increase the volume of concrete. In most work the aggregate should compose the greatest possible volume of the mix. A typical concrete mix consists of 75% aggregate and 25% paste. Proper aggregate selection according to size (gradation) assures maximum aggregate content in concrete.

A simple experiment can demonstrate this principle (Fig. 9-1). Fill a

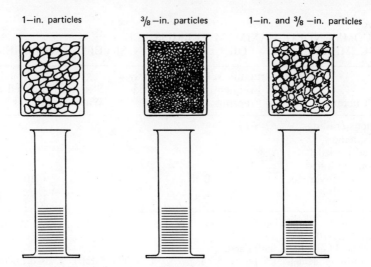

1—in. particles   3/8—in. particles   1—in. and 3/8—in. particles

Figure 9-1 The level of liquid in the graduates representing voids is constant if the particles are all the same size whether they are all large or all small. When different sizes are combined, or some large with some small, the void content decreases as illustrated by the graduate on the right.

glass with No. 50 sand (or any fine aggregate of uniform particle size). Pour water into the glass until it reaches the top; then pour it out and measure the quantity that was needed to fill the receptable. Repeat the procedure using 1-in. gravel (or any coarse aggregate of uniform particle size). The same amount of water is needed in both cases even though it would seem that the large spaces between the gravel would require much more water than the minute voids between the fine sand grains.

Combine the coarse and fine aggregate and repeat the procedure. Note that the amount of water needed to fill the glass is noticeably reduced. If other intermediate aggregate sizes are added, the quantity of water required to fill the voids diminishes progressively.

In concrete, cement paste, not water alone, fills the space between the aggregates. The lowest-cost mix will be the one with the best graded aggregates, that is, with the most complete range of sizes from extreme fines to maximum coarse, with no one size predominant.

By using the largest maximum coarse aggregate size permitted by form dimensions and reinforcement spacing, it is possible to incorporate the greatest variety of aggregate sizes into the concrete. The result for medium-strength mix designs is maximum economy. (As cement content increases beyond 500 lb per cu yd, aggregate gradation becomes less and less important.)

In addition to size and grading, several other characteristics affect the amount of paste required. For example, angular particles such as crushed rock require more paste to achieve good workability than do smooth rounded particles.

It is also important that fine aggregate have an adequate proportion of very small particles called *fines*. Concrete made with this type of fine aggregate is easier to finish than concrete made with coarse sand. The latter requires more cement paste to produce an equally acceptable surface finish.

ENTRAINED AIR

Intentionally entrained air in concrete has an effect on the proportions of materials used in the mix. The tiny bubbles of entrained air act as a lubricant between the aggregate particles causing them to slide over one another easily. The entrained air also reduces segregation. Aggregates have less tendency to settle, and there is very little bleed water. Because of this improvement in workability, air-entrained concrete may be mixed

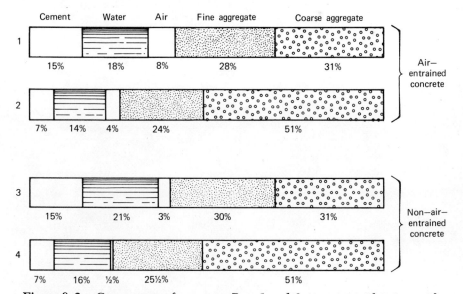

Figure 9–2 Components of concrete. Bars 1 and 3 represent rich mixes with small aggregates. Bars 2 and 4 represent lean mixes with large aggregates.

with less water, resulting in a lower water-cement ratio and a stronger paste. This compensates for the slight loss in strength which is characteristic of air entrainment in medium-strength mixes.

Figure 9-2 illustrates the range in proportions of materials used in concrete. Cement paste (cement, water, and sometimes entrained air) ordinarily constitutes 25 to 40% of the total volume of concrete. The absolute volume of cement is usually between 7 and 15% and that of water from 14 to 21%; aggregates constitute from 60 to 80%. Air contents of air-entrained concrete range up to about 8% of total concrete volume. Entrapped air will always be present at about 0.5 to 3.0% in non-air-entrained concrete.

CHAPTER 10
WORKABILITY

INTRODUCTION

Workability can be defined as the ease with which concrete can be mixed, transported, placed, and finished without loss of *homogeneity*. Different applications call for different degrees of workability. For example, concrete designed for pavement should be considerably stiffer than concrete designed for a heavily reinforced column. Concrete to be used in dams should be dryer than normal concrete; placing techniques would also vary.

Workability, from a placement standpoint, requires the elimination of as much entrapped air as possible; the concrete particles should be in maximum proximity to each other. This may be accomplished by vibration or ramming. Actually, the problem is to overcome friction between individual particles of concrete and between concrete and the surface of the mold or reinforcement. Workability is inversely proportional to the amount of labor needed to overcome this internal friction.

COMPACTION

Concrete must be workable enough to allow compaction to maximum density. Compaction is important because density is one of the factors directly related to strength. It eliminates inherent voids, bubbles, or fissures of entrapped air. As little as 5% voids in a volume of concrete can reduce strength as much as 39%, while 2% voids can reduce overall strength more than 10%.

The volume of excess water in concrete depends on the water-cement ratio of the mix. Water above the quantity needed for hydration evaporates, and leaves voids in the hardened concrete. Entrapped air (coarse voids not due to air entrainment) can be controlled by grading the fines in the aggregate.

SEGREGATION

Segregation is defined as the separation of the constituents of a heterogeneous concrete mixture so that their distribution is no longer uniform. Concrete is made of different-sized particles with varying specific gravities. These tend to settle at different rates if care is not taken to use suitably graded materials and to handle them with special care.

There are two patterns of segregation: first, the coarser particles tend to travel further along a slope or to settle more than finer particles; second, grout in wet mixes tends to separate from the aggregate.

Segregation, the separation of aggregates in concrete, is very detrimental to the final product. Every effort should be taken to avoid it. When segregation occurs, full compaction is impossible.

Hardened concrete with obvious imperfections is often the result of segregation. Some defects caused by segregation are rock pockets, sand streaking, weak and porous layers, crazing, and surface scaling. Since this type of damage is costly to repair, it is much better to avoid it by using well-designed mixes and by placing and finishing concrete properly.

Some occurrences of segregation are due to poor mix design, others to poor handling. Harsh mixes, extremely dry or undersanded mixes, may segregate even with careful handling, but segregation is more common in high slump concrete. On the other hand, even good concrete may segregate if mishandled. Methods of placing concrete that should be avoided are dropping and discharging against an obstacle or placing by using long, flat chutes. Special cohesive mixes must be designed for concrete to be placed by blowing down a long chute or through a pump. Segregation is likely to occur in concrete moved along inside forms or vibrated over a large area. There is always danger of segregation from improper or overlong use of vibrators.

Entrained air reduces the danger of segregation. Maintenance of a reasonable slump (between 1 and 4 in. for most applications) is helpful.

BLEEDING

Bleeding is a type of segregation in which mix water tends to rise to the surface of fresh concrete as solid constituents settle downward. It can produce porous, weak concrete subject to disintegration. When bleeding occurs in the top portion of a *lift*, that portion should be removed before another is placed. Bleed (excess) water may carry fine inert particles with it as it rises to the surface. The resultant material is called "laitance"; it prevents bonding between lifts (Fig. 10-1).

Finishing should not begin until the bleed water has evaporated or receded. If finishing is not delayed, a weak-wearing surface will develop. A wood float should be used and the surface should not be overworked during bleeding.

Bleed water becomes entrapped under coarse aggregate and reinforcement as it rises causing poor bond. This creates a weak horizontal plane in hardened concrete and invites damage.

Bleeding can be harmful. A porous surface is often the result of overworking concrete while bleed water is still present. When the water evap-

Figure 10–1 The puddles of water that have segregated out of the mix and have collected on top of this newly placed concrete are examples of bleeding.

orates, the cement and fine sand particles brought to the surface dry out causing dusting.

Bleeding can be controlled by proper mix design. No more water should be used than that required to produce a workable mix. Richer mixes and finely ground cements prevent water gain; smooth-working sands with adequate fines and well-graded aggregate are helpful. If it is not possible to get well-graded aggregate, it may be beneficial to use certain mineral admixtures (workability agents) to replace missing fine grain material. Air entrainment is another expedient. Water-reducing admixtures should be carefully tested before use since some of them tend to encourage bleeding.

EFFECT OF ADMIXTURES

Poor workability caused by badly shaped aggregate or poor gradation can be improved by the addition of an air-entraining agent; the agent provides reduced segregation, lower bleeding rate, and easier finishing. Air entrainment is helpful not only when the aggregate is less satisfactory; it is beneficial for the workability of concretes containing well-graded, well-shaped aggregate as well.

Lean, harsh mixes can be improved by the addition of fines; materials such as pozzolans, fly ash and rock dust improve plasticity. Small additions of bentonite or diatomaceous earth may also be helpful. Hydrated lime added to a lean mix (10 to 15% by weight of cement) will also increase workability. All of these, however, may be detrimental due to increased water demand. These materials should not be used in concrete with adequate fines, in rich mixes or, in any case, without prior testing to gauge their effect on strength and durability.

FACTORS AFFECTING WORKABILITY

Water

Water, and the amount used, affects the workability of concrete. A very dry or a very wet mix is hard to finish and presents problems in placement. Wet mixes tend to segregate, aggregate staying together in the center and cement paste running out to the edges of the placement area. Dry mixes do not vibrate or compact easily and require a tremendous amount of labor to finish. Water content can be reduced when air is entrained in the mix.

Aggregate

Aggregate gradation and water-cement ratio are interdependent at constant cement content. For each water-cement ratio there is one coarse-fine aggregate ratio (using given materials) that produces optimum workability. For a given workability there is one coarse-fine aggregate ratio which uses the lowest water content.

Aggregate of the rounded variety is most desirable for good workability. However, some harshness caused by elongated or flat pieces can be compensated for by using a richer mix with more cement and/or sand.

Sand grading should fall midpoint in the grading limit (Table 10-A). Sand that reaches the high limit on one sieve and the low limit on another should be avoided.

Coarse aggregate can vary more widely than sand in gradation without affecting workability noticeably. The amount of pea gravel ($\frac{3}{8}$- and $\frac{3}{16}$-in. sizes) has the greatest effect on workability, especially if it contains much undersized material.

Cement

Cement ground to the upper limits of fineness makes more workable concrete because it is more cohesive and has less tendency to segregate and bleed. Coarser cement, on the other hand, reduces the stickiness of the mix.

The amount of cement used in a mix has great effect on workability. A mix with very little cement (a lean mix) is harsh and difficult to finish. Good mix design can do much to eliminate this cause of poor workability.

The composition of the cement itself does not affect workability unless it contains a relatively large percentage of dehydrated gypsum, or plaster. This occurs when the temperature in the grinding mill is too high or the cement is stored in a high temperature. The result during mixing is a

Table 10-A
SAND GRADING LIMITS

Sieve Size	Cumulative Percent Passing
$\frac{3}{8}$ in.	100
No. 4	95–100
No. 8	75–90
No. 16	50–75
No. 30	30–60
No. 50	10–30
No. 100	2–8
Fineness modulus	2.50–3.00

phenomenon called "false set," "premature stiffening," "gun set," or "rubber set." Stiffening occurs soon after mixing during the dormant stage of hydration; it does not release any heat. This situation may be alleviated with long mixing or with remixing. On the job site, the difficulties of false set can be avoided by lengthening the mixing time. After remixing the concrete can be placed normally. Switching to cement ground at a lower temperature usually eliminates false set.

Flash set is different from false set. Flash set occurs when hot cement contacts the mixing water; it is actually accelerated hydration. Flash set can occur in winter when hot water is added to the cement in the mixer. It produces lumps of cement in which the center is dry cement surrounded with a shell of damp, partially hydrated cement.

This problem can be eliminated if hot water and cold aggregate are put in the mixer first, and cement is added after the aggregate has cooled the water.

EFFECTS OF TIME AND TEMPERATURE

Concrete stiffens because some of the water from the mix is absorbed by the aggregate, some is lost in evaporation, and some is combined in the hydration process. This affects workability. Ambient temperature also plays an important part. Hot days require the water content of a mix to be increased to maintain constant workability. As the temperature of the concrete is increased, the percentage of water required to effect a 1-in. change in slump also increases. Obviously, when concrete has to be hauled long distances in ready-mix trucks, high temperatures increase this water requirement.

Concrete should remain plastic for at least $\frac{1}{4}$ to $\frac{1}{2}$ hr after water is added without unusual precautions. Continuous agitation will maintain workability for 3 hr or more but may cause a significant slump loss. Water added to restore the original slump will often cause strength reductions.

MEASUREMENT OF WORKABILITY

Slump Test
The slump test is used very often in concrete work. It is easily performed at a job site and is useful in detecting variations in mixes of given proportions.

The slump test mold is the frustum of a cone 12 in. high. This is placed on a smooth, level surface with the smaller opening at the top. It is filled in three layers of equal volume, each of which is rodded 25 times with a rod having a hemispherical tip. The top is then struck off, and the cone is slowly lifted and set beside the unsupported concrete. The rod is laid across the cone and a measure of the distance from the bottom of the rod to the average top of the concrete is taken (Figs. 10-2 to 10-4).

A very stiff mix will have near zero slump. In such dry mixes it is hard to distinguish variations in workability. Lean mixes tend to be harsh and slumps can vary from true to shear (one side slides off the cone) in different samples of the same mix. The same slump can be recorded for concretes of different workabilities, depending on the aggregate used. The slump test is not a true determination of workability, but it is useful for on-site checks of variations in materials or mixing conditions.

The temperature of the concrete also affects the slump, as illustrated in Figure 10-5. A change in slump for a given batch amount of concrete in which the water content is constant gives the mixer operator a warning of change in the moisture content of the aggregate or change of aggregate gradation. Characteristics such as harshness and cohesiveness can be noted by studying the appearance of the slump specimen. Tapping the side of the fresh concrete specimen lightly with the rod causes a harsh, unworkable mix to crumble; cohesive, workable mixes stick together and appear to be unsegregated. A trained eye can judge concrete's workability merely by passing a trowel over the fresh sample. The effort

Figures 10-2, 10-3, 10-4   These pictures show the sequence followed in performing the slump test.

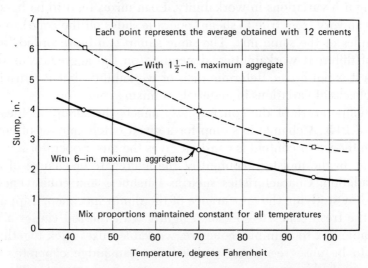

Figure 10–5 Relationship between slump and temperature of concrete made with two maximum sizes of aggregates. As the temperature of the mix ingredients increases, the slump decreases. *Concrete Manual,* Bureau of Reclamation.

needed to make a surface smooth and the appearance of that surface tells what kind of a mix has been prepared.

Flow Test
The flow test measures the spread of a pile of concrete subjected to jolting; its greatest value is in relation to segregation. It gives a good assessment of the consistency of stiff, rich, cohesive mixes. However, it is not normally performed in the field.

The flow test is described in ASTM C124. It is performed on a brass-top table 30 in. in diameter, mounted to drop ½ in. A mold, shaped like the frustum of a cone, is placed in the center of the table. It is filled with concrete in two equal, rodded layers. The mold is removed and the table jolted 15 times in about 15 sec. The concrete spreads and the average diameter of this spread is measured.

$$D = \text{average spread of concrete}$$
$$10 = \text{original base}$$
$$\text{flow } \% = \frac{D - 10}{10} \times 100$$

Values of 0 to 150% are attainable.

The flow test encourages and measures segregation; larger particles separate out and move to the edge of the table. A sloppy mix tends to run away from the center of the table leaving the coarser material behind. Workability is not gauged.

Remolding Test

The Powers remolding test also requires the use of a flow table. The test is performed in a vertical cylindrical vessel with a short tube or ring suspended in the middle which clears the bottom and sides of the vessel.

Figure 10–6 The Kelly ball test determines the consistency of a mix by measuring the depth that a 6 in. 30 lb ball will sink into freshly placed concrete.

A standard slump cone is placed in the center of the vessel and filled with concrete. When the slump cone is removed, the table is dropped. The number of drops required to bring the concrete to the same level on each side of the ring is recorded. This is called "remolding effort." Differences in the slump and remolding effort come from variations in the mix. In general, the greater the slump, the less effort needed to remold the concrete. The test is not suitable for field use.

Kelly Ball Test

The Kelly ball test determines the depth to which a 6-in.-diameter, 30-lb metal hemisphere will sink of its own weight into fresh concrete. It is a simple, quick test that can be done in the forms or in a wheelbarrow on a job site. The sample tested should be at least 8 in. deep and 18 in. wide. The test is used to measure variations in a mix due to changes in aggregate moisture content. When correlated to the slump test for a specific mix design, it can subsequently be used as a measure of slump. (See Fig. 10-6.)

CHAPTER 11
DURABILITY

INTRODUCTION

Durability of concrete refers to its ability to endure weathering action, attack by various chemical substances, freezing and thawing, abrasion, and many other conditions to which it may be exposed over the years. Aside from architectural potential, concrete is employed as a building material because of its strength and lasting qualities. These qualities are incorporated into concrete by the use of suitable materials, good mix proportioning, careful batching, mixing, handling, and placing, adequate consolidation, and sufficient curing.

Many of the conditions that cause concrete to lack durability are not immediately apparent. Some of them are harmful materials in the aggregate that cause cracking and surface blemishes, shrinkage of various aggregates that causes large deflections and cracking, use of highly absorptive aggregate that expands when moist and exerts sufficient force to disrupt concrete when frozen, and impure mixing water.

Poor concrete can be the result of improper construction practices as well as poor materials. Concrete that is too wet segregates easily during placing, as does well-proportioned concrete that is mishandled. Insufficient vibration causes porous or honeycombed concrete. Rain and hot or cold weather during construction can cause changes in concrete.

When concrete is placed in lifts it is essential that each lift be prepared properly before the next one is placed. Improper cleanup between lifts or at construction joints contributes to weak, permeable joints.

Reinforcing steel must be well embedded in durable concrete to protect it from corrosion. Quality concrete must be dense, made with sound aggregates and pure water, and mixed in proportions suitable for the planned use. It must be placed and cured with utmost care. Concrete handled in this way will give good service for many years.

WEATHERING

Weathering is defined as the inherent change of hardened concrete in terms of texture, strength, color, and other physical and chemical properties due to action of the elements.

All concrete weathers, but good concrete does so gradually and evenly. Gradual weathering shows up in rounded edges and/or slight roughening or erosion of the surface. This is a natural phenomenon that gives an interesting appearance to the concrete. Sound concrete that will break clean (through both the aggregate and paste) withstands weathering. Poor concrete that will break free of the aggregate and tend to be crumbly is subject to accelerated weathering.

Accelerated weathering shows up in cracks roughly parallel to the edges of the concrete. These cracks are known as D-line cracks. They fill with calcium carbonate and dirt. The concrete has a chalky, dull appearance and crumbles where aggregate separates from the paste. Given a blow with a hammer, poor concrete will produce a dead thump; good concrete, a ringing sound.

Sound concrete that will not be subjected to accelerated weathering depends on good mix proportioning, a low water-cement ratio, and good workmanship.

PERMEABILITY TO LIQUIDS

If concrete were completely solid, it might be watertight; however, air voids are always present. They make concrete permeable to outside influences. Materials in rain or ground water may penetrate concrete with adverse effects. Destruction from freezing and thawing in the presence of water, and corrosion of reinforcement subjected to moisture and air are problems related to the permeability of concrete. Permeable concrete is subject to movement of water from one surface to the other when there is a difference in the humidity on either surface.

Permeability is caused by pores in both the cement paste and the aggregate. Larger voids found in poor-quality concrete are caused by poor compaction and by bleeding. In very extreme cases, the voids may occupy as much as 10% of the volume of the concrete.

Fresh concrete is more permeable than hardened concrete. In fresh paste, the flow of water is controlled by the size, shape, and concentration of the original cement grains. Hydration reduces permeability as gel fills the original water-filled spaces. In mature paste, permeability depends on the size, shape, and concentration of gel particles and on whether or not the capillaries in the cement paste stop.

Permeability is lessened with a higher cement content in the concrete. This means less permeable concrete has a lower water-cement ratio.

Air entrainment reduces permeability because it reduces segregation and bleeding. At equal cement contents, the water-cement ratio is also reduced.

Permeability can be measured in a laboratory using the following test. The sides of a test specimen are sealed and water is applied under pressure to the top surface only. Compressed air can be used to apply the pressure. The quantity of water flowing through a given thickness of concrete within a given length of time is measured in cubic feet per second (cfs). Permeability is expressed as a coefficient of permeability with the following equation:

$$K_c = \frac{Q/A}{H/L}$$

where Q = rate of flow, cfs

A = area of cross section under pressure, square feet

H/L = ratio of head of water to percolation length

K_c = permeability coefficient, unit rate of discharge in cfs to a head of 1 ft acting on a specimen of 1 sq ft cross section and 1 ft thick

ATTACK BY SUBSTANCES

Acids

Some acids attack the cement paste in concrete causing disintegration at the surface or, if they enter at cracks, internal damage.

Some ores, coal, or cinders are sources of mineral acids when they become wet. Concrete used in contact with such materials must be protected. Unprotected concrete subjected to acid fumes in the atmosphere may be dissolved into a soft mass. This sometimes occurs in chimneys, steam railway tunnels, and other industrial situations.

Lactic, fatty, and acetic acids and blood found in food-processing plants, tanneries, and breweries attack concrete mildly but persistently. They cause a softening of concrete unless floors and walls are protected. These acids attack ceilings when they become volatilized.

Concrete in food-processing plants is attacked by bacteria and fungi that grow on floors and walls. These growths cause damage by mechanical action and by secretion of organic acids. Antibacterial cement is available; it decreases odors, slimes, and the rate of deterioration of concrete. The active ingredient in this cement is usually arsenic or copper. This type of cement is used in food plants, breweries, kitchens, dairies, and pharmaceutical plants as well as shower rooms, boathouses, and swimming pools.

Pure, flowing water formed by melting ice or condensation usually contains free carbon dioxide. In a water solution, carbon dioxide dissolves calcium hydroxide (a product of hydration) and causes erosion. Calcareous, rather than siliceous, aggregate gives best results under these circumstances.

Pure water containing no minerals can leach calcium hydroxide and other hydration products out of concrete. To combat this, the surface of concrete exposed to pure water should have good drainage and be smooth and dense.

Sulfates and Chlorides

Salts in solution can react with hardened cement paste. Ground water may contain dissolved mineral salts (alkalies), magnesium, and calcium sulfates. Sulfates react with calcium hydroxide and calcium aluminate hydrate. The products of this reaction (gypsum and calcium sulfoaluminate) require a greater volume than the compounds they replace thereby causing expansion and disruption of the concrete.

Sulfate attack produces a whitish deposit. Damage usually starts at the edges and corners, progressively spalling and cracking. Eventually the entire concrete mass is reduced to a soft state.

Calcium chloride is a commonly used deicing agent. When applied to ice or snow, it lowers the freezing point of the frozen material creating a salt solution that can again freeze. The effect of this freeze-thaw cycle on poor-quality and non-air-entrained concrete is scaling, spalling, and eventual failure. Proper air entrainment of concrete is ultimately the best protection against destruction from freezing and thawing cycles.

Chlorides accelerate rusting of iron and steel when exposed to air and moisture, a situation that could become worse if electrolytic action is present. When sulfate and chloride exposure is anticipated, a cement that is low in C_3A should be used, or pozzolans can be used to replace 15 to 30% (by weight) of the cement. Impermeability is important for resistance to sulfates and chlorides. This requires a rich mix with a low water-cement ratio.

High-pressure steam curing improves the resistance of concrete to sulfate attack. Both sulfate-resistant and ordinary portland cements can be used when sulfate attack is anticipated if the concrete is to be steam cured. The addition of calcium chloride to the mix reduces resistance to sulfates in all cements.

A test for sulfate resistance requires storing concrete specimens in a solution of sodium or magnesium sulfate or one of both. Alternate wetting and drying hastens damage. The loss of strength in the concrete is measured in terms of changes in dynamic modulus of elasticity as well as expansion.

Sewage
Sewage is not damaging in itself, but when the temperature is high, sulfur compounds contained in the effluvient can become reduced to hydrogen sulfide by anaerobic bacteria. When hydrogen sulfide is dissolved in moisture, this film forms on the exposed surfaces of concrete producing sulfuric acid strong enough to attack lime compounds in the concrete or cement. The condition can be discovered easily because it gives off a strong odor of hydrogen sulfide (similar to that of rotten eggs). Ventilation, to avoid formation of moisture, is helpful, but the best defense against damage is the use of a low water-cement ratio and carbonate aggregates.

Sea-Water
Sea water contains chemicals such as chlorides and sulfates. Crystallization of salts in the pores of concrete may cause disruption from the pressure exerted by the salt crystals. This takes place above the water level because crystallization occurs at the point of water evaporation. Permanently immersed concrete is attacked less than that which is alternately wetted and dried, and disintegration is more rapid in tropical

Table 11–A
EFFECT OF SUBSTANCES ON HARDENED CONCRETE

Group	Substance	Effect on Unprotected Concrete	Suggested Protection
A	Petroleum oils: heavy, light, or volatile	None, except some loss of oil from penetration of lighter oils	1, 2, 3, 4, 5, 6, or 7 (see list at end of the Table)
B	Coal tar distillates		
	Phenol, cresol, lysol, creosote	Slow attack	Same as above except 3 and 6
	Benzol, toluol, xylol, cumol	None except some loss of oil from penetration	Same as A, except 6, 7 for intermittent exposure
	Pitch, anthracene, carbonzol, paraffin	None	None required
C	Inorganic acids	Disintegration	
	Sulfurous		8, 9, or 10
	Sulfuric, nitric		8, 9, or 10 for concentration of 50% or less, below 150°F.
	Hydrochloric, hydrofluoric		9 or 10 for concentration of 50% or less, below 150°F.
D	Organic acids		
	Acetic	Slow disintegration	2, 4, 10, or 11
	Carbonic in water	Slow attack	1, 2, 4, 7, 11, or 14
	Lactic, tannic, citric	Slow attack	1, 2, 3, 4, 7, 11, 12, or 14
	Oxalic, dry carbonic	None	None
E	Organic oils	Very slight to slight	1, 2, 3, 7, or 13
F	Vegetable, fish, glycerine	Attack	
	Inorganic salts		
	Sulfates of aluminum, ammonia, magnesium, manganese, nickel, potassium, sodium, zinc	Active attack	1, 3, 7, 8, 10, 11, or 14
	Acid sulfate	Strong attack	Same

190

Group	Substance	Effect on Unprotected Concrete	Suggested Protection
	Ammonium sulfate and nitrate	Strong attack	Same
	Chlorides of aluminum, copper, iron, magnesium, mercury	Slight attack	Same
G	Chlorides of calcium, potassium, sodium, strontium	No attack	None
	Nitrates of calcium, potassium, sodium		
	Soluble sulfides (except ammonia)		
	Carbonates, fluorides, silicates		
H (misc.)	Bromine	Active attack	8
	Carbon bisulfide	Impure vapor attacks	8
	Coke	In presence of water forms H_2SO_4	See group C
	Hydrogen sulfide		7
	Brine	If continuously wet	None
		Otherwise, use	3, 7

Protective treatments
1. Fluorosilicate, sodium silicate.
2. Spar varnish.
3. Linseed oil.
4. Phenol formaldehyde varnish.
5. Thiokol, Amercoat.
6. Prestressed concrete.
7. Epoxy resin surface sealer.
8. Glass, vitrified brick, or tile laid in litharge.
9. Lead.
10. Rubber.
11. Bituminous paint or enamel.
12. Paraffin.
13. Bakelite varnish.
14. Neoprene solution in solvent and resin.

J. J. Waddell, *Practical Quality Control for Concrete.*

climates than in temperate regions. Since damage can occur only when water permeates concrete, impermeability is most important. Chemical damage is further aggravated by the action of sea water, frost, and wave impact.

To protect against sea-water attack, concrete should have well-graded, first-class, nonreactive aggregates. Air should be entrained to increase workability and slump in a low water-cement ratio mix. A minimum of 665 lb of cement per cubic yard of concrete produces a good workable mix for this type of exposure. Table 11-A lists some of the substances that concrete comes in contact with, their effects on concrete, and the possible ways to protect against attack.

ABRASION

Resistance to abrasion (wear) is important with many uses of concrete. Among these uses are floors, pavements, and spillways. Strong concrete, naturally, is more resistant to abrasion than weak concrete (Fig. 11-1). Wear of concrete surfaces is most often due to the following:

1. Attrition—wear on concrete floors due to normal foot traffic, light trucking, and sliding objects
2. Attrition combined with scraping and impact—wear on concrete road surfaces due to studded tires and vehicles with and without chains

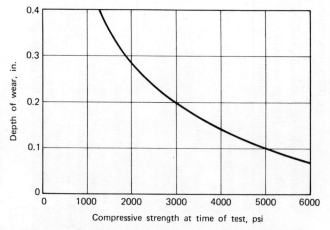

Figure 11-1   This graph shows the effect of compressive strength resistance of concrete. High-strength concrete is highly resistant to abrasion.

3. Attrition and scraping—wear on underwater construction due to the action of abrasive materials carried by moving water
4. Impact—wear on hydraulic structures where a high hydraulic gradient is present.

Concrete designed to resist abrasion should have a low water-cement ratio and a minimum of fine aggregate. It must be properly placed and cured, and overworking of the surface must be prevented to avoid excess sand and water rising to the top.

Abrasion resistance increases greatly as the compressive strength of concrete increases to about 6000 spi. The effect of entrained air on abrasion resistance is comparable to its effect on strength. Where abrasion is likely, air entrainment should be limited if freeze-thaw exposure permits. As an alternate, the water-cement ratio should be lowered.

FREEZE-THAW

Fresh Concrete

Occasionally, fresh, plastic concrete freezes during initial hardening. This should be carefully avoided because it reduces durability, weather resistance, and strength as much as one half. How long fresh concrete remains frozen is not important. Once it has been frozen, it may never attain its full potential, despite long curing at a reasonable temperature.

Hardened Concrete

In some climates, hardened concrete is subjected to freezing and thawing cycles. Concrete without entrained air can be damaged when water in the pores freezes and expands, causing cracks. This is a cumulative process. Each new cold cycle freezes water in existing cracks, enlarging them and producing new cracks.

Resistance to frost depends on the strength of the cement paste and its pore structure. Concrete that is thoroughly dried before freezing is more frost resistant. The further along the hydration process has progressed, the more frost resistant the concrete will be since more hydration means less free water is available to freeze and higher tensile strength is possible (Fig. 11-2).

If possible, aggregates used in concrete that will be exposed to freezing and thawing should be thoroughly dried before use so no moisture remains to freeze and expand. Saturated aggregates contribute to the destruction of concrete.

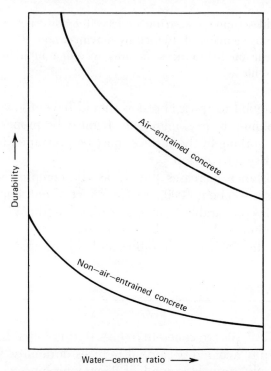

Figure 11–2 Idealized relationship between freeze-thaw resistance and water-cement ratio for air-entrained and non-air-entrained concretes. High resistance to freezing and thawing is associated with entrained air and low water-cement ratio.

Deicing salts are commonly used in climates where concrete is subjected to freezing. They cause osmotic pressure that tends to resist the movement of water away from an advancing ice front, thereby increasing disruptive action. Good concrete resists the action of calcium chloride or sodium chloride but not the action of all salts. Deicing salts with ammonium sulfate or ammonium nitrate should be avoided because they attack the concrete chemically. No deicing salts should be applied the first winter after placement.

ENTRAINED AIR

Concrete generally has approximately 1.0 to 3.0% entrapped air. This air is ineffective in reducing resistance to freezing and thawing because the voids are too large and are prone to become filled with hydration products. Intentionally entrained air has much smaller bubbles that are separated by cement paste. There are fewer channels incorporated in the paste through which water can pass. Since the voids are filled with air and hydration can only take place in water, the voids may never fill with products of hydration (Fig. 11-3).

The size and spacing of the air bubbles is important. Entrained air is closely spaced and evenly distributed; entrapped air bubbles are randomly distributed. Entrained air bubbles range in size from 0.001 to 0.003 in. in diameter; entrapped air voids can be as large as 0.5 in. in diameter. There are approximately 5 billion bubbles in 1 cu yd of non-air-entrained concrete, while there may be as many as 300 to 500 billion in air-entrained concrete (seven sacks of cement per cu yd, 7% entrained air with 3/8-in. maximum-sized aggregate).

Additions of 4 to 6% of entrained air, with an aggregate of 1- 1/2-in. maximum size, greatly increases resistance to freeze-thaw cycles. As the maximum size of coarse aggregate decreases, the amount of entrained air

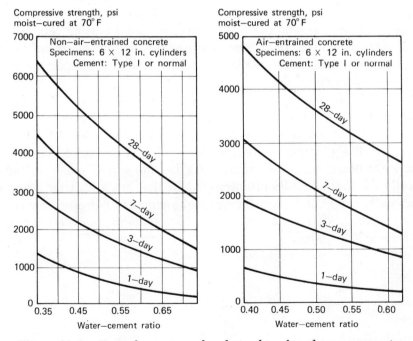

Figure 11–3 Typical age-strength relationships based on compression tests of 6 × 12-in. cylinders (right, for air-entrained concrete; left, for non-air-entrained concrete).

needed for the same effect increases. A maximum size aggregate of ¼ in. should have 8 to 10% air content.

Air entrainment lowers strength, but since less water is needed to produce concrete of the same workability, the actual effect may sometimes increase strength, particularly in lean mixes. Rich mixes (richer than five sacks of cement per cubic yards) usually suffer strength losses.

Large amounts of small air bubbles serve as reservoirs for relief of pressure developed within freezing concrete. Water in the concrete is forced into the entrained air voids; hydraulic pressure is relieved and high-tensile stresses that lead to failure are avoided. During thawing, water returns from the voids to the cement paste, and the cycle is ready to be repeated.

In general, the more air-entraining agent added to a mix, the more air entrained. This is true up to a certain point beyond which there is no increase in the volume of voids. Other factors influencing the amount of air entrained for a given amount of air-entraining agent are (1) workability of concrete—more workable concrete holds more air than drier mixes; (2) fineness of cement—finer cement decreases the effectiveness of air entraining; (3) aggregate gradation—volume of air decreases when an excess of very fine sand particles is used, but sand passing Nos. 30 and 50 and retained on No. 100 sieves increases the amount of air entrained; and (4) aggregate shape—angular rather than round aggregate increases the amount of air entrained.

Mixing operations also affect the amount of entrained air in a mix. If the mixing time is too short, the agent will not be dispersed. With too much time, some air is lost. High temperatures cause loss of air. (An increase of 40°F in temperature from 50° to 90°F, for example, halves the net amount of air actually entrained.) Vibration expels some air, but it is mostly the large bubbles of entrapped air that are lost. After 3 min of vibration, however, only half the original amount of entrained air is left in the concrete. After 9 min, as little as 20% remains, so overvibrating is obviously to be avoided.

CHAPTER 12
STRENGTH OF CONCRETE

INTRODUCTION

The potential strength of concrete is determined by the properties and proportions of its constituent materials. Strict control of all materials and processes is essential to maintain concrete production with properties constant from batch to batch.

All commercially marketed portland cements meet minimum industry specified property standards. Different cements, however, do have varying properties. Better concrete performance is always achieved by proportioning the mixture to take full advantage of the properties of the cement to be used on the job.

Aggregate also varies from batch to batch. Ideally, one source of aggregate should be used for an entire project to ensure consistent quality. If this is not possible, it is important to run tests using the new aggregate and possibly to redesign the mix to assure adequate quality for the job. Variations in aggregate grading may call for changes in the amount of mix water; this may affect properties of the concrete. Undesirable materials in the aggregate—clay lumps, soft particles, organic matter, silt, mica, lignite, and easily friable pieces—should be at a minimum.

Although they detract from workability, cuboidal aggregates are preferable where strength is the main consideration. Elongated and flat aggregates should be avoided because of the possibility that water can become entrapped under them causing poor bond and cracking when it freezes. They also require more water for equal workability.

Water itself should not contain harmful materials such as sugar, silt, oils, acids, alkalis and their salts, organic material, or sewage. If water from a certain source has been used successfully in making concrete, it may be used again. If not, concrete tests using the water in question should be made. These tests should include determination of setting time, strength, and soundness.

Admixtures may sometimes be used to modify properties of concrete, but they should not be expected to compensate for improper proportioning or poor workmanship. The amounts used should be constantly checked because small changes in quantity may seriously affect the properties of the concrete. Also, various combinations of admixtures, when several are mixed, may cause adverse effects if ingredients combine to form new substances.

Strength is affected by the water-cement ratio, the age of the concrete, and the quality of curing as well as the size and shape of the specimen tested. A specimen tested for compressive strength will have a higher indicated strength if the specimen is dry before testing, whereas flexural strength will be lower in a dry specimen. Strength must be carefully planned, designed, and controlled.

MATERIALS AFFECTING STRENGTH

Cement

Modern cements are made under close quality control and seldom cause poor concrete if handled properly. Among reasons why cement might be at fault in poor concrete are the presence of moisture and carbon dioxide contamination caused by improper sealing of bags or excessive storage time. Good cement is floury and contains no lumps that cannot be broken easily with the fingers.

Different types of cement have different strength-producing characteristics. The raw materials used by cement plants vary as do the methods of processing them. Fineness and finish-grinding temperature of cement affect concrete strength. Finer grinding increases cement strength development, especially at early ages.

Probably the most important factor determining the ultimate strength of concrete is the water-cement ratio. A low ratio of water to cement will produce stronger concrete; water in excess of that needed for hydration tends to separate the cement particles and leave voids in the concrete.

High-early-strength cement (Type III) produces concrete with greater strength at 7, 14, and 28 days than do the other (Fig. 12-1). At 90 days all five types develop comparable strengths. If continually fog cured at 70°F for five years, the high-early-strength cement will have developed less strength than the others, with Type V showing the greatest strength. The reasons for the development of strength early with Type III cement are that it contains an increased percentage of tricalcium silicate and/or tricalcium aluminate and is more finely ground than normal cement.

Dehydration of the gypsum in cement, caused by too high a temperature in the finish grinding, can cause premature stiffening in freshly

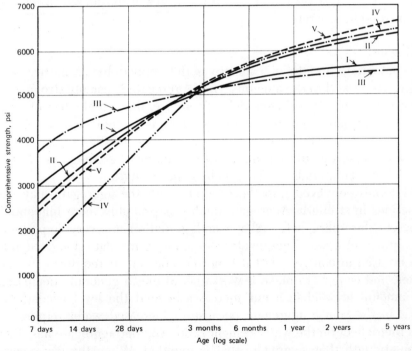

Figure 12-1   Rates of strength development for concrete made with various types of cement. Each test batch was made to these specifications: 6 sacks of cement per cubic yard: aggregate 0 to 1½ in. in 6 × 12-in. cylinders: fog-cured at 70°F for the ages shown. *Concrete Manual,* U.S. Bureau of Reclamation.

mixed concrete, a phenomenon called "false" or "rubber" set. If more mixing water is added, workability is restored, but a loss in strength results due to the increase in water-cement ratio. An alternative means of restoring workability is continued mixing, since false set generally occurs about 60 sec after initial mixing is started. This can be accomplished without the attendant loss in the strength potential of the concrete.

Water

If there is any doubt as to the quality of the mix water to be used in concrete, it should be tested in mortar cubes. If the tests result in 7- and 28-day strengths equal to at least 90% of specimens made with water that is known to be satisfactory in concrete, and if the setting time is not adversely affected, the water can be used. Stagnant pools or swamps should not be considered sources, especially if they contain moss or algae.

Strength deficiency can be caused by as little as 0.2% organic material in mix water. Sugar is extremely detrimental to setting, and salt can corrode reinforcing steel. In general, if the water is considered drinkable, it can be used in concrete.

Aggregate

The potential strength of concrete is determined by properties of the aggregate as well as the amounts used. Aggregate is usually stronger than the cement paste that surrounds it. Therefore, the actual effect of normal weight aggregate is not of great importance to strength unless very high strengths are specified. However, some characteristics of aggregate—particle shape, texture, maximum size, soundness, grading, and freedom from harmful materials—affect bonding and strength.

Materials conveyed into the concrete with the aggregate may cause variations in strength. Aggregate quality is probably more influential to concrete durability, while shape and gradation are very important to strength. Well-graded aggregate containing many flat, elongated pieces can be used in concrete, but it is not economical. It requires extra sand, water, and cement to make it workable. Strength generally decreases as the amount of sand in a mix increases beyond the level needed to fill voids in the coarse aggregate because of increased water requirements and their effect on the water-cement ratio. Angular aggregate has slightly more strength than aggregate that is rounded. When the water-cement ratio of the paste is constant, the compressive strength of concrete varies with the size of the coarse aggregate. Aggregate of 3/4- to 1-in. maximum size has been found to produce the strongest concrete with moderate cement contents.

Aggregate should meet the requirements of ASTM C33. Soft, friable aggregate should be avoided. Aggregate with a very low specific gravity or excessive absorption (except lightweight aggregate) should not be used. If tests and case histories of aggregate from a specific source are satisfactory, that aggregate can safely be used.

Aggregate should be washed to free it from loam, clay, rock dust, or other coatings that have adverse effects on strength. Loose fines in small amounts are not harmful.

Organic matter such as humus is harmful. When present in surface soils of an aggregate deposite, the surface layer should be stripped off to a great enough depth to rid the aggregate of this material.

Slag aggregate, unless it is properly prepared, can be a source of poor strength. If it has a high sulfate content, it can cause poor strength and even deterioration. Excess sulfate in slag reacts with the hydrated calcium aluminate of cement to cause harmful expansion.

Admixtures

The effect of admixtures on concrete strength depends on the properties of the admixture and the characteristics of the concrete mix. There are admixtures that are designed to serve in many capacities—water reducers, accelerators, and air entrainers, for example. Admixtures vary from producer to producer; only reputable products should be used.

Calcium chloride (Fig. 12-2) is one of the most commonly used. It increases early strength during cold weather and protects concrete from damage due to freezing by shortening the time needed for hardening; however, it should never be considered an "antifreeze" for concrete. The amount of calcium chloride used should be limited to that necessary to produce the desired results. In any case, it should never exceed 2% by weight of cement.

Calcium chloride can have undesirable effects such as shrinkage and warping, reduced resistance to sulfate attack, reduced resistance to alkali-aggregate reaction, and corrosion of reinforcing steel. It is best used during freezing weather to produce high early strength and to shorten time of set; it is most effective in rich mixes. Air-entraining agents should be added to the mix separately from the calcium chloride or they will be precipitated. They can be added to the sand while the calcium chloride is flowed in with the water.

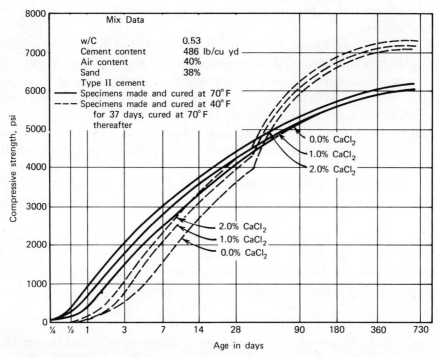

Figure 12-2 Effect of small additions of calcium chloride on the compressive strength of concrete. *Concrete Manual*, U.S. Bureau of Reclamation.

Some organic compounds are used to reduce water requirements or to retard set. Decreased water content produces reduced water-cement ratio that causes increased strength and decreased permeability.

Often 5 to 7½% entrained air is needed to produce good workability and durability. This has the effect of reducing strength. With the water-cement ratio constant, concrete compressive strength falls about 5% for each 1% of air that is entrained. Keeping the cement content constant and reducing the water-cement ratio for a given workability reduces strength losses and may even increase strength.

Some of the finely divided mineral admixtures on the market are (1) relatively chemically inert materials (ground quartz, ground limestone, bentonite, hydrated lime, and talc); (2) cementitious materials (natural cements, hydraulic limes, slag cements, and granulated blast-furnace slag); and (3) pozzolans, including materials such as fly ash, volcanic glass, diatomaceous earth, and some shales and clays. These are added to improve resistance to sulfates, acids, and the like, and to increase impermeability. Strength changes occur with the use of these admixtures. Lean mixtures are apt to improve while rich mixes may be slightly affected in terms of strength losses. The changes are slow, especially at low temperatures. Cementitious and pozzolanic admixtures give higher strengths at later ages when they are continuously moist cured. Fly ash, up to 30% by weight of cement, reduces compressive strength at 7 and 28 days but may increase it after three months.

An admixture such as a retarder can delay the setting of concrete. Early strength gains may also be delayed by the use of selective retarders. Retarders are often used when hot weather causes early setting and reduction in slump due to higher concrete temperatures. An overdose of retarder can cause low early strength, but if not too severe, this lag can be regained through continued moist curing.

PRODUCTION

Weighing and measuring the ingredients of concrete and the methods used to introduce these ingredients in the mixer are collectively called batching. Variations in concrete strength are produced by these batching procedures. Careful attention must be paid to all factors involved. Proportions must be correct. Consideration of the amount of water in the aggregate must be made when determining the amount of water to be added. Losses of materials, especially cement and water, must be prevented. A written record of all quantities in each batch should be kept as a check for on-the-job control and for possible future reference in the event of unexpected results.

Inaccuracies in measuring materials may be due to volumetric batching, inaccurate scales, careless weighing procedures, materials sticking to the weigh hopper, and uncompensated variations in aggregate moisture content. Also, variations in the sequence in which materials are charged into the mixer cause changes in strength.

Proper mixing procedures and length of mixing time are important. Mixing must be long enough to produce a homogeneous mix, but no longer than necessary. Overmixing reduces slump and causes an increase in fines from the grinding action of the materials. Delays in transporting and placing also cause slump loss. If water is added to the mix to bring back original slump, strength is lost. Concrete mixed in winter and not placed within 2 hr should be discarded; in summer, allowable placement time is even smaller.

Segregation must be avoided during handling and placing. Concrete varies in quality from one part to another if segregation occurs. Equal compaction and consolidation is necessary for uniform strength and quality.

Proper curing (Fig. 12-3) is essential to quality concrete. Otherwise excellent concrete can be ruined if improperly cured. Concrete will continue to gain strength as long as moisture is available and temperature is satisfactory; loss of moisture or excessively low temperatures will stop strength gain.

An external air temperature of 80°F or above, during and just after placing, will result in lower ultimate strength, as compared to concrete placed at 40° to 80°F. Low-curing temperatures cause slow strength development, but full strength can be achieved if proper curing is continued long enough.

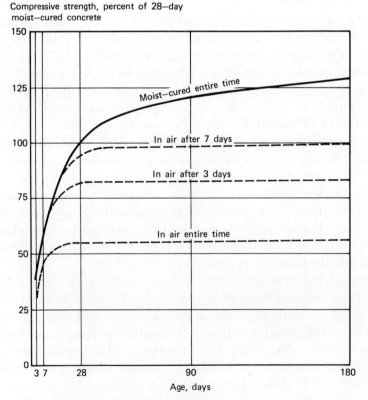

Figure 12-3 Strength of concrete continues to increase as long as moisture is present for hydration of cement.

Freezing immediately after mixing may reduce strength up to 50%. After 500 psi compressive strength is attained, concrete may be considered resistant to early freezing.

Compressive strength increases at a decreasing rate as the moist curing period increases. Strength development may stop at an early age if concrete is allowed to dry. When moist curing is stopped, compressive strength increases for a short time but soon stops. When moist curing is resumed after a period of drying, strength increases are also resumed unless the concrete is too impermeable.

Steam curing, either low or high pressure, is an excellent means of moist curing. Low-pressure steam curing usually employs saturated steam at temperatures from 135° to 195°F. The greatest acceleration in strength gain and the minimum loss in ultimate strength is effected at 130° to 165°F. High-pressure steam curing (autoclave) is performed at a maximum steam pressure of 80 to 170 psi, about 325° to 375°F. Concrete cured in this manner attains in 24 hr approximately the same strength as concrete moist cured at room temperature does in 28 days (Fig. 12-4).

Heat is released during the reaction between portland cement and water. It can be used to elevate the temperature of the concrete to accelerate its strength gain. If the area is properly insulated, this heat helps cure concrete in cold weather. Under steam-curing conditions, the heat

Figure 12–4 This small autoclave is used at the Portland Cement Association's laboratories to cure concrete masonry units under high-stream pressure. The units are used for test purposes.

of hydration serves to raise the concrete temperature the same as in normal curing. Once concrete reaches the ambient temperature, the air temperature outside the concrete should be reduced slowly to allow gradual cooling.

If the temperature of the concrete is allowed to greatly exceed the temperature in the enclosure, moisture may be lost unless steam is used as a source of heat. Concrete should remain at normal temperatures for 2 to 4 hr after casting before it is subjected to high temperatures. It can be damaged by removal from forms too rapidly or by being left in forms too long. Depending on the use for which a concrete is designed, forms should be removed only when the concrete reaches satisfactory strengths determined by previously designed strength tests.

WATER-CEMENT RATIO

A basic law of concrete technology states that the strength of concrete within the plastic range is inversely proportional to the water-cement ratio, all other variables remaining the same. In other words, the less water used for a certain amount of cement, the stronger the concrete will be.

Different exposure conditions require different water-cement ratios. But under some conditions the water-cement ratio should be selected on the basis of required strength. In such cases, if at all possible, tests to determine the relationship between water-cement ratio and strength should be made with actual job materials. If laboratory test data or experience records for this relationship cannot be obtained because of time limitations, the necessary water-cement ratio may be estimated from mix design tables.

If flexural strength rather than compressive strength is the basis for design—as in pavements—tests should be made to determine the relationship between water-cement ratio and flexural strength. An approximate relationship between flexural and compressive strength is:

$$f'_c \cong \left(\frac{R}{K}\right)$$

in which f'_c = compressive strength, in pounds per square inch
  R = flexural strength (modulus of rupture), in pounds per square inch, third-point loading
  K = a constant, usually between 8 and 10

When both exposure conditions and strength must be considered, the lower of these two indicated water-cement ratios should be used.

TESTING FOR STRENGTH

Results of testing concrete for strength are subject to many variables. Concrete is made from different materials, by different methods, and in different mix proportions and is tested by different procedures. Actually, these are all reasons why concrete must be tested in a regular pattern during a job rather than just once at the beginning.

The usual procedure is to cast test specimens as a structure is placed. The measure of strength shown by the test specimen is not a true picture of the concrete's strength; test specimens are cured differently from the structure and so do not reflect the workmanship that goes into the structure. The test is valuable as an approximation of actual strength, a test of potential quality as delivered. Such tests keep batches of concrete uniform throughout a job.

Compressive Strength

Concrete is usually proportioned for a given compressive strength (resistance to axial loading) at a given age. The test of compressive strength is the one most frequently used. Cylinders, 6 in. in diameter and 12 in. long, are cast, or cores are taken and tested at specified ages. Test specimens may show different results depending on their moisture content, the loading rate of the test apparatus, the test temperature, lateral restraint, and many other variables.

Compressive test specimens are placed in an apparatus that applies force on both ends until failure; this test is performed in a laboratory (Fig. 12-5).

Tensile Strength

Cracking occurs when contraction due to chemical activity, drying, shrinkage, or a decrease in temperature is restrained. (Tests of tensile strength are important because concrete by nature has poor tensile qualities.) Direct tensile tests are seldom used, but "splitting tensile strength" is determined by applying increasing compressive loads on 6 × 12-in. cylinders placed on their sides and compressed until splitting occurs. Tensile strength is then calculated from the maximum load and the known dimensions. The procedure, described in ASTM C496, is

$$T = \frac{2P}{ld}$$

where T = splitting tensile strength, in pounds per square inch
              P = maximum applied load, in pounds
              l = length, in inches
              d = diameter, in inches

Flexural Strength

The ability to withstand bending is known as flexural strength. Bending tests are made on unreinforced concrete beams subjected to flexural loading (highway pavements, for example). The modulus of rupture is calculated from a test in which the beams are supported at both ends and

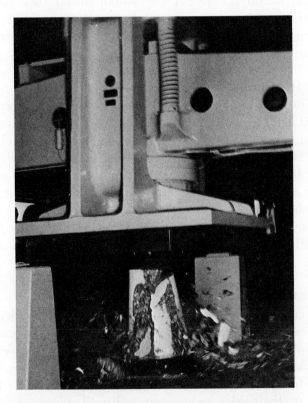

Figure 12-5   The test cylinder collapses at the point where it can no longer support the tremendous pressure applied by the press. Measurement of that point determines compressive strength.

loaded with a concentrated force at the center. This determines the strength of small fine concrete specimens or, with the concentrated forces at one-third and two-thirds points, larger specimens. Stress on the fiber farthest from the neutral axis is calculated.

Beams for flexural tests are 6×6-in. cross sections and are tested over a span *three times their depth* for highway use. These tests are common for mix design and job control.

Bonding Strength

ASTM C234 describes procedures used to test bonding between reinforcing steel and concrete. The test is used mainly for research since it cannot be performed in the field.

Bonding strength is tested by the "pullout" test. Strength is noted at the initial slip of reinforcement at the free end and at a slip of 0.01 in. at the loaded end. This type of strength is influenced by the type of reinforcement. Deformed bars have much better bonding ability than plain reinforcing. Bond strength varies approximately with compressive strength below 3000 psi. Above 3000 psi, the increase in bond stress continues, but more slowly, up to 6000 psi.

The quality of cement paste and the amount of air entrainment are significant factors in bonding strength. Delayed vibration, properly timed and applied, increases bond strength. It is lower for horizontal bars than for vertical ones because water collects on the underside of the former, interfering with bond.

Modulus of Elasticity

Modulus of elasticity (Young's modulus) is a measure of stiffness or resistance to deformation in hardened concrete. It normally varies between 2 and 6 million psi, depending on the compressive strength. The same factors that cause strength increases also cause increases in the modulus of elasticity.

Poisson's Ratio

Poisson's ratio is the ratio of lateral strain to longitudinal strain within the elastic range for axially loaded specimens. The ratio is required for structural analysis and design in many types of structures. The procedure for determining the ratio is described in ASTM C469. Poisson's ratio probably increases with age up to about 1 year. It falls within a range of 0.10 to 0.20 at an age of approximately 50 years. When experimental data is not available, it is assumed to be approximately $1/6$.

Fatigue Strength

Fatigue failure is caused by repeated additions of loads smaller than any single static load that could cause failure at a comparable level. The fatigue limit is the stress level below which a specimen of concrete will withstand an infinite number of cyles of load application without failure.

Many conditions govern the fatigue limit; they include type of loading, type of concrete (plain, reinforced, prestressed), strength of concrete, and ambient conditions, for example.

Stress: The intensity of the force developed within a body resisting forces acting on it.

Strain: The deformation of a material subjected to forces expressed as a ratio of linear unit changes to the distance within which that movement occurs.

General Considerations

Test specimens may be standard cured (moist $73.5° \pm 3°F$), or they may be cured in the field under the same conditions as the structure. Field-cured specimens are usually tested to determine when forms may be removed and when the structure may be used.

The number of specimens tested for a given structure is often specified. Tests may be specified on a given volume of concrete, for a given area of paved surface, or for a certain number of tests a day. Three specimens are normally required for each test. Tests are generally made at 7 and 28 days, but may be required for 30, 60, or 90 days or, in some cases, for as much as a year.

Nondestructive Tests

Nondestructive tests are used on specimens and structures to determine the properties of the concrete in place. As described in the *Concrete Construction Handbook*, they are:

1. Indentation tests—in which the rebound of a spring-driven hammer is measured and related to ultimate strength.
2. Sonic tests—usually involving determination of the resonant frequency of longitudinal, transverse, or torsional vibration of small concrete specimens. These tests provide information on dynamic modulus of elasticity (the ratio of normal stress to corresponding strain).
3. Pulse-transmission tests at sonic and ultrasonic frequencies—tests that measure the velocity of a compressional pulse traveling through the concrete, and provide information on the presence or absence of

cracking in monolithic concrete. They may also be used to measure the thickness of slabs on inaccessible faces.

4. Radioactive tests involve absorption of gamma and X-rays—tests that provide information on density or quality of concrete and also on the presence or absence of reinforcing steel.

The most widely used rebound device is the Schmidt concrete test hammer. Though not a conclusive testing device, it supplements information derived from other tests. It provides rapid results and is simple and economical to use. The results are subject to many variables: the position of the hammer; the smoothness of the test surface; internal and surface moisture content; size, shape, and rigidity of the specimen; aggregate types, sizes, and concentration near the surface; air pockets; age; temperature; and previous curing conditions of the specimen. Many of these variables can be overcome by calibrating the hammer for specific existing conditions.

The "resistance to penetration" test, in frequent use today, measures the relationship between compressive strength of concrete and the energy or force required to set steel test probes into hardened concrete. One procedure is called the Windsor Probe Test System, which includes a powder activated probe driver and probes. These devices are set first by firing a driver, then measuring the exposed heights of probes after their entry into the hardened concrete. The probe system, like the hammer system, can be used to correlate estimates of hardened concrete compressive strengths but cannot provide precise quantitative measurements.

Sonic frequency tests are used for lab testing of beams and cylinders. They follow the course of deterioration of specimens subjected to weathering or exposure.

STRENGTH RELATIONSHIPS

Compressive, shearing, tensile, flexural, and bond strengths are all interrelated. An increase or decrease in one is reflected in the others, but not always to the same degree. Tensile strength usually ranges between 8 and 12% of compressive strength, averaging about 10%. The modulus of rupture is usually between 12 and 20% of compressive strength, averaging 15% for concrete with a compressive strength of 3500 psi.

Splitting tensile strength averages 67% of flexural strength for concrete made with crushed stone, 62% for concrete made with gravel, and 76% for concrete made with lightweight aggregate. Splitting tensile strength averages 11% of compressive strength for concrete made with crushed stone and gravel, and 8% for lightweight concrete.

Concrete cylinders, tested in axial compression, usually fail by shearing along an inclined plane. This reaction is due to a combination of tensile and shearing stresses along the plane. Shearing stresses cause tensile and compressive stresses, all closely related. Concrete is weaker in tension than in shear, so failure occurs basically as a result of tensile stresses. Testing for shearing stress is difficult, but it appears that shearing strength is about 20% of compressive strength.

SECTION FOUR
PROPORTIONING AND MIXING CONCRETE

CHAPTER 13
DESIGN OF CONCRETE MIXTURES

INTRODUCTION

The purpose of proportioning is to make economical concrete that is satisfactorily workable and has all of the specified qualities after hardening. In mixing most compounds, the proportions of the various ingredients can be varied without seriously affecting the product. Coffee and water can be varied somewhat to produce a weaker or stronger (but nonetheless acceptable) brew. The same is true of paint, cake batter, or glue. In concrete, however, the amounts of the various constituents in the mix have a critical effect on the qualities and usefulness of the final product. As a result, before proportioning can begin, it is necessary to know what kind of end product is desired, and what performance is expected. Concrete for a bridge deck would not have the same qualities as concrete for a massive dam; concrete designed for use in Arizona may not be satisfactory when exposed to the temperature extremes found in Canada.

The strength required of the concrete must be known, as well as anticipated exposure conditions, availability and costs of component materials, amount of reinforcing steel or prestress cables to be used, and special requirements for transporting, placing, and finishing. All these requirements and limitations must be considered before the proportions of materials can be calculated, regardless of planned mix-design methods. When the use, exposure conditions, and engineer's or architect's specifications for required strength are known, mix characteristics—water-cement ratio, slump, entrained air content, types and proportion of aggregates, and pounds of cement per cubic yard of concrete—can be determined.

ARBITRARY PROPORTIONS

In the past, concrete was proportioned by volume, and strength was controlled by varying the cement content. For example, a rich mix was 1:1:2 (1 part cement, 1 part sand, 2 parts gravel); a lean mix, 1:3:6 (1 part cement, 3 parts sand, 6 parts gravel). An experienced man who knew his materials could often produce good results with this method. However, even an experienced man would occasionally fail because this arbitrary volume method disregarded the water content of the aggregates. In addition, if aggregates were from different sources or if the gradation, shape, or surface texture of the aggregates varied, the characteristics of the resulting concrete would change. Arbitrary proportioning is now limited to small jobs where suitable proportions have been established by experience or observation.

PROPORTIONING BY MAXIMUM DENSITY OF AGGREGATES

The economy of a concrete mix is dependent on the use of minimum cement-water paste combined with maximum proportions of coarse and fine aggregates; this results in maximum density and minimum volume of voids. Experimentation with percentages of fine and coarse aggregates permits the development of a curve such as illustrated by Fig. 13-1.

In this curve, the point of greatest aggregate density (38% of sand by weight) is used. However, the optimum percentage of sand, which requires a minimum amount of paste, is almost always below this percentage for maximum density of aggregate. When the paste is introduced into the mix of aggregates, it separates the aggregates because it, too, occupies space. As a result, the method of proportioning by maximum density of aggregates is not used today. This type of density curve is, however, often used to help in selecting satisfactory gradations of aggregates.

PROPORTIONING BY SURFACE AREA OF AGGREGATES

Since the amount of cement paste required in concrete is affected by the amount of surface to be coated, the surface area of the aggregates seems

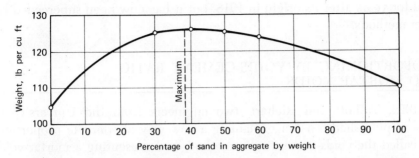

Figure 13-1   Unit weight of dry-rodded fine and coarse aggregates mixed in various proportions. (For illustration only, to show trends: values will differ with aggregates.) Drawing from *Composition and Properties of Concrete*, Troxell, Davis, and Kelly.

to be a logical basis for proportioning. This method does not, however, consider the fact that the amount of paste required in concrete is also affected by the number of voids to be filled. As a result, the method of proportioning by surface area of aggregates is much too inaccurate.

PROPORTIONING BY FINENESS MODULUS BY AGGREGATE

Concurrent with the development of the water-cement ratio basis for proportioning, a method of establishing fine and coarse proportions based on the fineness modulus evolved. The fineness modulus is an index number roughly proportional to the average particle size of a given aggregate; that is, the coarser the aggregate, the greater the fineness modulus. It is computed by adding the cumulative percentages coarser than the Nos. 4, 8, 16, 30, 50, and 100 U.S. standard sieves, and dividing the sum by 100. Although the fineness modulus gives no idea of gradation and does not distinguish between a single size aggregate and a graded aggregate having the same average size, it is useful in indicating whether one graded aggregate sample is finer or coarser than another.

Charts were developed to show the relationship between slump, fineness modulus, volume of aggregate, volume of cement, strength, and maximum size of aggregate. When the slump, fineness modulus, and maximum aggregate size were known, the compressive strength and volume of mixed aggregate could be read from the chart. With this information, the percentage of sand in the mix could then be computed using another chart. The fineness-modulus method of proportioning was widely used for years after its origin in 1918, but it has now been superceded by other methods.

PROPORTIONING BY VOIDS-CEMENT RATIO AND MORTAR VOIDS

In 1922, Talbott and Richart, two engineers from the University of Illinois, published a paper explaining a new way of concrete proportioning called the voids-cement ratio method, and presenting a mortar-voids system for applying this method. Spaces not occupied by aggregate or cement were considered voids; their volume, the sum of the mix's water and entrapped air. The voids-cement ratio, expressed in cubic feet of voids per hundredweight (cwt), is approximately proportional to the water-cement ratio because of the relatively insignificant amount of space occupied by air in concrete.

To produce concrete of a particular strength, a voids-cement ratio for this strength must be determined. This process involves preparing a number of trial mortar mixes. Several batches are blended, each with a different ratio of sand to cement. Water is added gradually to each, and the volume of voids for each different proportion of sand, cement, and water is determined. This step is accomplished by weighing a known volume of mortar, using the specific gravities of the constituents of the mortar and the known weight and volume of the sample produced, calculating the volume of voids. Each resulting sample is subsequently tested for strength.

This experimentation produces raw data from which a number of curves can be developed. For each water content, four different curves can be drawn: (1) the relationship of strength to the voids-cement ratio, (2) the voids-cement ratio to the sand-cement ratio, (3) the sand-cement ratio to the voids-per-unit volume of mortar, and (4) the sand-cement ratio to the water-per-unit volume of mortar. Once these curves have been developed, a final estimate of proportions can be made.

First, it is necessary to assume, from previous experience, the amount of water needed to produce the desired consistency or slump in the concrete. Referring to the series for that water content, proportioning with the amount of strength desired in the concrete can begin. Knowing this, determination of the voids-cement ratio from curve No. 1 is possible. Once the voids-cement ratio is established, the sand-cement ratio can be determined from curve No. 2. With the sand-cement ratio known, the voids-per-unit volume of mortar can be determined from curve No. 3, and the water-per-unit volume of mortar from curve No. 4. At this point, the amount of water-per-unit volume of mortar and the sand-cement ratio are known, and the volume of sand- and cement-per-unit volume of mortar figures can be established. The only unknown remaining is the volume of coarse aggregate to be used in a unit volume of concrete. This amount can only be arrived at by trial or from past experience. The voids-cement-ratio method of proportioning is complicated and involves the preparation of a large number of trial mixes and numerous computations. Even after all this effort, the quantities of water and coarse aggregate are still subject to trial or experience. For these reasons, the voids-cement ratio and the mortar-voids methods of proportioning are used very little outside of the laboratory.

The research done by Talbott and Richart in developing this method, however, increased the general knowledge of concrete qualities and behavior. For instance, the voids-cement ratio to strength curves explain air-entrained concrete's lower strength, despite controlled, constant water-cement ratio.

Another insight resulting from the Talbott and Richart experiments was knowledge of the way the volume of a sample of mortar changes as water, in small increments, is added. Initially, volume increases because of the presence of surface films on cement and sand particles. This causes "bulking." Additional water causes volume to decrease by breaking down water added. The water content at which voids are just filled with water surface films and filling voids with water. Further additions of water cause volume of the mixture to increase proportionately to the amount of and the volume of the mortar is at a minimum is called the basic water content. The amount of water required for workability is always greater than this basic water content.

PROPORTIONING BY VOID CEMENT OF COARSE AGGREGATE

In 1942, National Crushed Stone Association researchers Goldbeck and Gray, building on the original work of Talbott and Richart, published a method of proportioning based on void content of coarse aggregate. Although this method is no longer used, it included a table of recommended bulk volumes of coarse aggregate per unit volume of concrete when the fineness modulus of sand and the maximum size of coarse aggregate is known. The table (Table 13-A), applicable to all cement contents and to crushed or rounded aggregates, was later incorporated into the ACI as the absolute volume method of proportioning.

Table 13-A
VOLUME OF COARSE AGGREGATE PER UNIT VOLUME OF CONCRETE[a]

Maximum Size of Aggregate, in.	Volume of Dry-Rodded Coarse Aggregate per Unit Volume of Concrete for Different Fineness Moduli of Sand			
	2.40	2.60	2.80	3.00
⅜	0.50	0.48	0.46	0.44
½	0.59	0.57	0.55	0.53
¾	0.66	0.64	0.62	0.60
1	0.71	0.69	0.67	0.65
1½	0.75	0.73	0.71	0.69
2	0.78	0.76	0.74	0.72
3	0.82	0.80	0.78	0.76
6	0.87	0.85	0.83	0.81

[a] ACI 211. These volumes are selected from empirical relationships to produce concrete with a degree of workability suitable for usual reinforced construction. For less-workable concrete, such as required for pavement construction, they may be increased about 10%. When placement is to be by pump, they should be reduced about 10%.

COMPOSITION AND PROPERTIES OF CONCRETE

Before the proportions of concrete can be determined, it is necessary to know the intended use of the concrete, its exposure conditions, and what strength, durability, and special characteristics it must possess. Since most of the desired properties of hardened concrete are dependent on the quality of the cement paste, the first step in proportioning is establishing the paste's water-cement ratio. Once it has been established, gradation and maximum size of the aggregates and consistency of the concrete can be varied widely without strength or other desirable qualities being affected. If, however, the water-cement ratio is altered by adding water, the resulting concrete can no longer adequately serve its end purpose. For this reason, an appropriate water-cement ratio must be selected and maintained throughout an entire job, or that portion of it having the same requirements.

In determining the water-cement ratio, four factors must be considered: (1) the type of structure in which the concrete will be used, (2) the exposure conditions, (3) the contact, if any, with sulfates in soils and ground waters, and (4) the strength required of the structure. Once these are known, the water-cement ratio can be estimated through the use of time-tested tables. Table 13-B gives recommendations for water-cement ratios for the two basic types of structures and for two degrees of severe exposure. Figure 13-2 provides recommendations for water-cement ratios

Table 13–B
MAXIMUM PERMISSIBLE WATER-CEMENT RATIOS
FOR CONCRETE IN SEVERE EXPOSURES[a]

Type of Structure	Structure Wet Continuously or Frequently and Exposed to Freezing and Thawing	Structure Exposed to Sea Water or Sulfates
Thin sections (railings, curbs, sills, ledges, ornamental work) and sections with less than 1 in. cover over steel	0.45[b]	0.40[c]
All other structures	0.50[b]	0.45[c]

[a] Based on report of ACI Committee 201, "Durability of Concrete in Service," previously cited.
[b] Concrete should also be air entrained.
[c] If sulfate-resisting cement (Type II or Type V of ASTM C150) is used, permissible water-cement ratio may be increased by 0.05.

needed to achieve various predetermined strengths. The lower of these recommendations will specify the ratio to be used.

When the water-cement ratio has been established, there are still a number of factors that must be considered before final mix proportions are selected. One of these is slump range, an indication of the consistency of the concrete. Typical slump ranges for various types of construction are listed in Table 13-C.

A good aggregate for concrete is evenly graded, from the smallest particle of sand to the largest of coarse aggregate. The larger this maximum size, the more economical the resulting concrete will be. There is a practical limit, however, to maximum aggregate size. Generally, it should not exceed one-fifth of the minimum dimension of the concrete member nor three-fourths of the clear space between reinforcing bars or between reinforcement and forms. For unreinforced slabs on the ground, the maximum size should not exceed one-third the thickness of slab.

Air entrainment should be utilized in any concrete exposed to freezing and thawing; it may also be used to improve workability. Air-entrained concrete can be made by using an air-entrained portland cement, by adding an air-entraining admixture during mixing, or by a combination of both methods. Air-entraining cements will generally provide an adequate amount of air to meet most job conditions. If an admixture is used, it can be readily adjusted to meet various job conditions. The amount of admixture recommended by the manufacturer will usually produce the desired air content. Recommended total air contents for air-entrained concrete

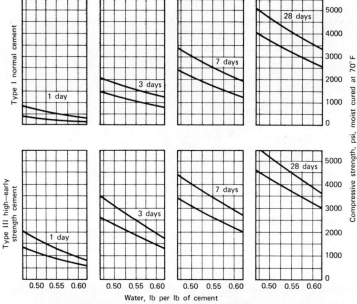

*Concrete with air content within recommended limits and maximum aggregate size of 2 in. or less

Non—air—entrained
concrete

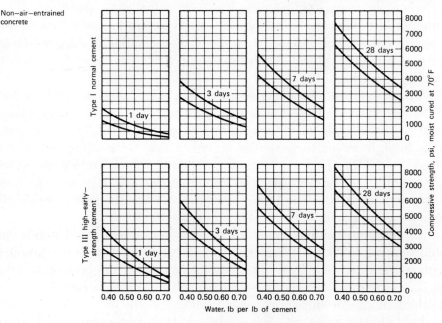

Figure 13–2 Relationship between water-cement ratio and compressive strength for portland cements at different ages. Lower curves give lower limit of range; upper curves, upper limit of range. These relationships are approximate and should be used only as a guide in lieu of data on job materials.

Table 13–C
TYPICAL SLUMP RANGES FOR VARIOUS TYPES OF CONSTRUCTION

Types of Construction	Slump, in.	
	Maximum[a]	Minimum
Reinforced foundation walls and footings	3	1
Unreinforced footings, caissons, and substructure walls	3	1
Reinforced slabs, beams, and walls	4	1
Building columns	4	1
Pavements and slabs	3	1
Heavy mass construction	2	1

[a] May be increased 1 in. for methods of consolidation other than vibration.

are shown in Table 13-D. Note that the amount of entrained air required to produce adequate freeze-thaw resistance is dependent on maximum aggregate size. Entrained air is contained in the mortar portion of concrete. In properly proportioned mixes, mortar content and, therefore, entrained air decrease as maximum aggregate size increases. Air-entraining agents generally reduce the amount of mixing water required as well. The resulting decrease in water-cement ratio helps offset possible strength loss and results in improved durability. The strength of air-entrained concrete is equal or nearly equal to that of non-air-entrained concrete having identical cement content and slump.

In some cases, an architect's or engineer's specifications require that concrete have a minimum cement content; this assures concrete of adequate durability and strength regardless of any aggregate gradation or water content variations. Minimum cement requirements also ensure the suitable finishability of slabs and quality of vertical surfaces where little inspection is expected.

PROPORTIONING METHODS

Although there are numerous methods of proportioning, the two most widely used are the Unit-Weight Trial Mix Method and the Absolute Volume Method. Regardless of the method chosen, the objectives of the proportioning—desired characteristics of the concrete in both the plastic and hardened states—must be known.

In fresh concrete, workability is the most important characteristic; it is a function of cohesiveness, plasticity, and consistency. Unfortunately, there is no completely accurate way of measuring workability. A number of instruments are, however, effectively used for this purpose. The slump

Table 13-D
APPROXIMATE MIXING WATER REQUIREMENTS AND CONTENT REQUIREMENTS FOR DIFFERENT SLUMPS AND MAXIMUM SIZES OF AGGREGATES[a]

Maximum Size of Aggregate, in.	Air-Entrained Concrete				Non-Air-Entrained Concrete			
	Recommended Average Total Air Content, percent	Slump, in.[b]			Approximate Amount of Entrapped Air, percent	Slump, in.		
		Water, lb per cu yd of Concrete[c]				Water, lb per cu yd of Concrete[c]		
		1 to 2	3 to 4	6 to 7		1 to 2	3 to 4	6 to 7
3/8	8	305	340	365	3	350	385	410
1/2	7	295	325	345	2.5	335	365	385
3/4	6	280	305	325	2	315	340	360
1	5	270	295	310	1.5	300	325	340
1½	4.5	250	275	290	1	275	300	315
2	4	240	265	280	0.5	260	285	300
3	3.5	225	250	270	0.3	240	265	285

[a] Adapted from Recommended Practice for Selecting Proportions for Normal Weight Concrete (ACI 211.1-70).
[b] The slump values for concrete containing aggregate larger than 1½ in. are based on slump tests made after removal of particles larger than 1½ in. by wet-screening.
[c] These quantities of mixing water are for use in computing cement factors for trial batches. They are maximums for reasonable well-shaped angular coarse aggregates graded within limits of accepted specifications.

cone, a simple, economical, and portable piece of equipment, and the probe (ACI, Vol. LXIV, No. 5, p. 270), basically a hollow tube that permits easy passage of both paste and aggregate, are the most widely used. Other common devices include the penetration ball, the flow table, and the Vebe apparatus. These devices do not actually measure workability, a qualitative measurement, but instead they measure consistency, often used as a measure of workability. Workability is affected by all mix factors, that is, water and cement content, aggregate size, gradation, and texture, and amount of entrained air. Often, harsh mixes, considered to have poor workability, can be readily placed in the field through proper handling, placing, finishing, and adequate vibration.

In hardened concrete, durability and strength are the properties most often needed. High density, watertightness, and a high modulus of elasticity are often necessary, as are good thermal properties, low volume changes, and low creep. Most of these properties will be present if the concrete has adequate initial strength and durability. Durability refers to resistance to weathering—freezing and thawing, wetting and drying, temperature fluctuations—and resistance to erosion or wear, impact, internal and external chemical attack, and water penetration. A durable concrete is obtainable through good form work, adequate consolidation, low water-cement ratio, air entrainment, and proper curing and finishing techniques.

High quality materials are essential to the production of durable concrete. The end product of any mix design is desirable concrete in both the plastic and, most importantly, the hardened state. Such concrete is achieved by proper cement content or water-cement ratio, maximum size of aggregate, air content, and desired workability, as measured by slump. When these characteristics have been determined, proportioning can begin.

The Unit-Weight Method

The unit-weight method of proportioning is a good way to determine a mix design when very little is known about the materials to be used. It involves very few calculations and makes use of saturated, surface-dry aggregates making it unnecessary to determine aggregate moisture contents, specific gravity, and fineness modulus.

When the quality of a concrete mix is specified in terms of water-cement ratio, the unit-weight method includes two basic steps: mixing a cement paste to the required specifications and insuring that the resulting concrete will have the required strength and durability.

Fine and coarse aggregates are blended into the cement paste until required slump and workability are obtained. The weight of each ingredient used in the trial mix is then calculated, and from these weights, the quantities of ingredients required per cubic yard of concrete are projected. Representative samples of aggregate, cement, water, and air-entraining admixtures must be used. To simplify calculations and to eliminate sources of error caused by variations in aggregate-moisture content, the aggregates should be presoaked and allowed to dry to a saturated, surface-dry condition. The aggregates are then placed in covered containers to maintain SSD condition until they are used.

The size of the trial batch is dependent on the mixing equipment available and on the number and size of test specimens to be made. If batches are to be mixed by hand and if no test specimens are required, a 10-lb batch may be adequate. However, larger batches produce more accurate data, and machine mixing is recommended because it more nearly represents job conditions and is mandatory for air-entrained concrete. A detailed explanation of the unit-weight method, with sample problems and data sheets, follows.

Trial Mixes Using the Unit-Weight Method of Mix Design. The most direct method for determining optimum mix proportions is through the use of trial mixes based on the unit-weight method of mix design. There is no specific limit to the size of a batch, and such mixes may be relatively small, laboratory-precise batches or job-size batches made during the course of normal concrete production. In either case, the batch requirements (water-cement ratio, maximum size of aggregate, air content, and slump) must be selected before proportions can be computed.

A series of mixes is made by maintaining each of these requirements constant while varying the amounts of fine and coarse aggregates. Proportions of the components are then computed for each mix, and the results of the trial mixes are compared. Based on economic considerations and the workability of the trial mixes, final mix proportions can be selected. One advantage in using this method of mix design is that the technician does not need specific laboratory data on the materials; it is only necessary that the materials used be representative of those to be used in the actual project, and that the aggregate be in a saturated, surface-dry condition.

Problem Project Design Data. Air-entrained concrete is required for a retaining wall in Detroit, Michigan. It will be exposed to air on one side and fresh water on the other. The engineer's specifications require a minimum structural strength of 3000 psi at 28 days. From the contract drawings, it is determined that the minimum wall thickness is to be 12 in. and that 2 in. of concrete are required to cover the reinforcing steel.

Type I cement is to be used by the ready mixed concrete producer. High-frequency vibrators are to be used to consolidate the concrete.

The technician enters this information on the first trial-mix data sheet. (See Data Sheet 13-a, below.) Then, through the use of various charts and tables (all of which were covered earlier in this chapter), he determines the basic mix design data as follows (Data Sheet 13-A):

Water-cement ratio. According to Table 13-B (since Detroit, Michigan, is located in a severe exposure area), the technician notes that this structure does not fit the descriptions given in the first category, placing it among "all other structures." And since the concrete is to be exposed to fresh water, the required water-cement ratio for durability is 0.50 lb of water per pound of cement.

The structural requirements call for a strength of 3000 psi at 28 days. From Figure 13-2 (air-entrained concrete, Type I cement), to ensure a strength of 3000* psi would require a water-cement ratio of, at most, 0.57. Since the lower water-cement ratio should govern the mix design, the ratio of 0.50 for durability will hold, even though it will produce strength higher than required for structural purposes.

Aggregate size. From the information given, the maximum size of coarse aggregate should not exceed $1\frac{1}{2}$ in., which is three-fourths of the distance between the reinforcing steel and the forms ($\frac{3}{4} \times 2$ in.). If this size aggregate is not economically available, the next smaller size should be specified.

Air-content. Because of the severe exposure conditions and the engineer's specifications, the concrete should be air entrained. From Table 13-D, he obtains the required air content based on the maximum size of the aggregate. In this case, for $1\frac{1}{2}$ in. aggregate, the air content should average 4.5%. This amount of air entrainment is usually achieved using an air-entraining admixture or Type IA cement. The admixture manufacturer's recommendations as to dosage should be followed.

---

* To ensure that the concrete used in the structure will produce the required strength, however, the strength of the laboratory mix design would usually be increased by a predetermined factor. This step is omitted in the examples. ACI Standard Recommended Practice for Evaluation of Compressive Test Results of Field Concrete (ACI 214) recommends that the strength of this mix design be increased by 15% (3000 psi×1.15=3450 psi). With this strength, 2 test cylinders out of every 10 would be expected to have less than the required 3000 psi.

Data Sheet 13–A
CONCRETE MIX DESIGN—UNIT WEIGHT METHOD
Specifications

JOB LOCATION: *Detroit, Michigan*

CONCRETE TO BE USED FOR: *Retaining wall exposed to fresh water*

INFORMATION AND SKETCH OF CONCRETE:

EXPOSURE CONDITIONS:

Mild _____ Severe __✓__

In Air _____ In Fresh Water __✓__

In Sea Water or in Contact with Sulfates _____

MAXIMUM W/C RATIO FOR EXPOSURE: __0.50__ lb water/lb cement

SPECIFIED STRENGTH: __3000__ psi at __28__ days

MAXIMUM W/C RATIO FOR STRENGTH: __0.57__ lb water/lb cement

W/C RATIO TO USE: __0.50__ lb water/lb cement

MAXIMUM AGGREGATE SIZE: __1½__ in.

AIR CONTENT: __4.5__ % ± %

RECOMMENDED SLUMP:

Minimum: __1__ in. Maximum: __3__ in. Use: __2__ in.

Slump. The slump range for various types of construction is obtained from Table 13-C. Since a retaining wall is a reinforced foundation wall and vibration is specified, the slump range should be between 1 and 3 in.

At this stage in the mix design, all data on the first sheet has been filled in and the actual mix design, using saturated, surface-dry aggregate, can begin.

Batch weights. For convenience, a batch containing 10 lb of cement will be prepared. Since the required water-cement ratio is 0.50 lb of water per pound of cement, 5 lb of water will be used. Cement and water are placed in a mixer with the required amount of air-entraining agent and mixed. An arbitrary volume of fine and coarse aggregate, that is enough to supply the needed quantity for the trial mix, 60 and 50 lb respectively, is weighed. These initial weights are entered in column 2 of the table on the second data sheet (see Data Sheet 13-B, below). The weights for cement and water are also recorded in column 4.

The aggregates are added to the cement paste until a workable concrete mixture, with adequate slump for placement in a retaining wall, is obtained. An experienced technician can determine if the slump is adequate, and can also observe if the concrete is sufficiently workable. He controls slump by adding fine and coarse aggregates in small increments.

Excessive amounts of coarse aggregate will produce a harsh, unworkable mix; excessive amounts of fine aggregate will result in an uneconomical mix that is also somewhat unworkable with high cement and water contents and greater volume changes. A good rule of thumb is to use as much coarse aggregate as possible, and fine aggregate sufficient to ensure adequate workability. Mechanical mixing, approximating the mixing used on the actual project, should be used to obtain the most reliable trial mix results.

If the slump is greater than required, small quantities of fine or coarse aggregates (or both) are gradually added until the required slump is reached. If the slump is less than that required, water and cement in the appropriate ratio are added to produce the desired slump. With the latter procedure, it is important that the amounts of water and cement be measured accurately and noted on the data sheet.

When the slump is within the recommended range, its value is recorded on the data sheet and a test for air content is made. Air content is also noted on the data sheet along with a description of the concrete's workability (in this case, good) and appearance (sandy). Final weights of fine and coarse aggregates are recorded in column 3 along with the corresponding weights used in column 4.

CONCRETE MIX DESIGN—UNIT-WEIGHT METHOD
Date and Calculations for Trial Batch _#1_
(Saturated, surface-dry aggregates)

Batch size: 10 lb ✓ 20 lb ——— 50 lb ——— 100 lb ——— of cement

Instructions:
- a. Complete columns 2, 3, and 4 of table.
- b. Fill in items below table.
- c. Do calculations 1-5 at bottom.
- d. Complete column 5 using _N_ from calculation 5.

(1) Material	(2) Initial Wt (lb.)	(3) Final Wt (lb) (4)	Wt Used (Col. 2-Col. 3) (lb)	(5) Wt per Cu Yd of Concrete	
				N × Col. 4 (lb/cu yd)	Remarks
Cement	10.0	0	10.0	508	Cement Factor
Water	5.0	0	5.0	254	Water Content
Fine Aggregate	60.0	35.0	25.0	1270 (F)	$\% \ FA^a = \dfrac{\text{wt of FA}}{\text{wt of agg.}} \times 100$
Coarse Aggregate	50.0	12.4	37.6	1910 (G)	$= \dfrac{F}{F+G} \times 100$ $= \dfrac{1270}{3180} \times 100 = \underline{40}\%$
Totals			77.6 (T)	3942	Total Wt
			N × T = 77.6 × 60.8 = 3942		Math check: should equal total wt, above ✓

Calculations. Using a cylinder of known weight and volume, a test for unit weight is made. The cylinder is filled with a sample of concrete and then weighed. Net weight of concrete (without the container) is determined at line 1 of the data sheet, and unit weight of concrete (net weight of concrete divided by the volume of the container) is calculated at line 2.

Knowing the unit weight, it is then possible to compute (a) the volume of concrete per batch (line 3), (b) the yield (or volume) per 100 lb of cement (line 4), and (c) the number of batches per cubic yard (line 5). Once the number of batches per cubic yard has been established, it is possible to calculate the cubic yard weights of all materials and the percentage of fine aggregate, as shown in column 5 of the table. When all this is done, one trial mix is completed.

Data Sheet 13-B (Continued)

Amount of Air-entraining Admixture Used: __0.3__ oz

Measured Air Content: __5.3__ % Measured Slump: __2¼__ in.

Appearance: Sandy __✓__ Good _____ Rocky _____

Workability: Excellent _____ Good __✓__ Fair _____ Poor _____

Wt of container and concrete (A):   __43.5__ lb

Wt of container (B):   __7.0__ lb

Vol of container (C):   __0.25__ cu ft

(1) Weight of concrete (W) = A − B =   __36.5__ lb

(2) Unit weight of concrete (UW) = $\dfrac{\text{wt of concrete}}{\text{vol of container}}$

$$= \frac{W}{C} = \frac{36.5}{0.25} = \qquad\qquad \underline{146.0}\ \text{lb/cu ft}$$

(3) Volume of concrete (V) = $\dfrac{\text{total wt of material used}}{\text{unit wt of concrete}}$

$$= \frac{T}{UW} = \frac{77.6}{146} = \qquad\qquad \underline{0.532}\ \text{cu ft}$$

(4) Yield = vol of concrete per 100 lb of cement

$$= V \times \frac{100}{\text{batch size}} = 0.532 \times \frac{100}{10} = \qquad \underline{5.32}\ \text{cu ft}$$

(5) Number of Batches (__10__ lb of cement each) per

cu yd of concrete (N) = $\dfrac{27\ \text{cu ft/cu yd}}{\text{vol of concrete}} = \dfrac{}{0.532}$ __50.8__ batches/cu yd

---

a Percent fine aggregate of total aggregate.

To determine the most economical proportions, additional trial batches should be made varying the percentage of fine aggregate, and taking due regard of the limits of workability. In each batch, the water-cement ratio, aggregate gradation, air content, and slump are maintained as constant as possible.

UNIT-WEIGHT METHOD OF MIX DESIGN— SECOND TRIAL BATCH

In the previous problem, a mix was designed for a specific reinforced-concrete retaininig wall. In this problem, a second mix will be designed for the same project varying the percentage of sand. Cement and water contents of 10 lb and 5 lb, respectively, will remain the same. In order to

Data Sheet 13–C
CONCRETE MIX DESIGN—UNIT-WEIGHT METHOD
Date and Caculations for Trial Batch __#2__
(Saturated, surface-dry aggregates)

Batch size: 10 lb __✓__ 20 lb _____ 50 lb _____ 100 lb _____ of cement

Instructions:
   a. Complete columns 2, 3, and 4 of table.
   b. Fill in items below table.
   c. Do calculations 1-5 at bottom.
   d. Complete column 5 using N for calculation 5.

(1) Material	(2) Initial Wt (lb)	(3) Final Wt (lb)	(4) Wt Used (Col. 2- Col. 3) (lb)	(5) Wt per Cu Yd of Concrete — N × Col. 4 (lb/cu yd)	(5) Wt per Cu Yd of Concrete — Remarks
Cement	10.0	0	10.0	511	Cement Factor
Water	5.0	0	5.0	255	Water Content
Fine Aggregate	18.8	0	18.8	961 (F)	$\% \text{FA}^a = \dfrac{\text{Wt of FA}}{\text{Wt of Agg}} \times 100$
Coarse Aggregate	60.0	16.3	43.7	2233 (G)	$= \dfrac{F}{F+G} \times 100$ $= \dfrac{961}{3194} \times 100 = \underline{\ 30\ } \%$
Totals			77.5(T)	3960	Total Wt
			N × T = 77.5 ×51.1 =3960		Math check: should equal total wt, above

change the sand percentage, the sand content of the previous mix will be reduced by 6 to 7 lb. Data Sheet -C for a typical example is given below. The following are some of the results obtained: a sand content of 30% of total aggregate weight, a cement content of 511 lb per cu yd, and a yield of 5.28 cu ft per 100 lb of cement.

OPTIMUM PROPORTIONS FOR ECONOMY

In the two previous problems, trial mixes were presented covering differ-ent percentages of fine aggregate. To determine the most economical

Data Sheet 13-C (Continued)

Amount of Air-entraining Admixture Used: __0.3__ oz

Measured Air Content: __5.2__ %   Measured Slump: __3__ in.

Appearance: Sandy _____ Good _____ Rocky __✓__

Workability: Excellent _____ Good _____ Fair _____ Poor __✓__

Wt of container and concrete (A):   __43.7__ lb

Wt of container (B):   __7.0__ lb

Vol of container (C):   __0.25__ cu ft

(1) Weight of concrete (W) = A − B =   __36.7__ lb

(2) Unit Weight of concrete (UW) = $\dfrac{\text{wt of concrete}}{\text{vol of container}}$

$= \dfrac{W}{C} = \dfrac{36.7}{0.25} =$ __146.8__ lb/cu ft

(3) Volume of concrete (V) = $\dfrac{\text{Total wt of material used}}{\text{unit wt of concrete}}$

$= \dfrac{T}{UW} = \dfrac{77.5}{146.8} =$ __0.528__ cu ft

(4) Yield = vol. of concrete per 100 lb of cement

$= V \times \dfrac{100}{\text{batch size}} = 0.528 \times \dfrac{100}{10} =$ __5.28__ cu ft

(5) Number of Batches (__10__ lb of cement each) per cu yd of

concrete (N) $= \dfrac{27 \text{ cu ft/cu yd}}{\text{vol of concrete}} = \dfrac{27}{V}\ \dfrac{27}{0.528} =$ __51.1__ batches/cu yd

---

a Percent fine aggregate of total aggregate.

proportions, additional trial batches should be made further varying the fine aggregate percentage. As in the second example, water-cement ratio, aggregate gradation, air content, and slump should be maintained as constant as possible. Two additional examples have been worked, and all four mixes are tabulated in Table 13-E. The relationship between cement content (pound per cubic yard) and percentage fine aggregate (% of total aggregate) is plotted in Figure 13-3. For these specific materials, the minimum cement factor (bottom of the curve) occurs at a sand percentage of 36%.

Since the water-cement ratio for this mix is 0.50 and the unit weight is

Table 13-E
EXAMPLE RESULTS OF LABORATORY TRIAL MIXES

Batch Number	Slump in.	Percent Air	Unit Weight lb/cu ft	Cement Content lb/cu yd	Percent Fine Aggregate	Yield cu ft/cwt	Workability
1	2¼	5.3	146.0	508	40	5.32	Good
2	3	5.2	146.8	511	30	5.28	Harsh
3	2½	5.1	146.0	501	35	5.39	Good
4	2¾	5.3	146.0	503	38	5.37	Excellent

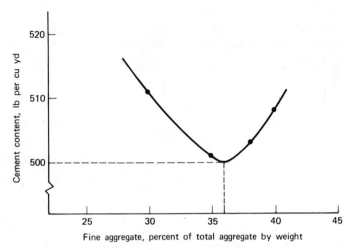

Figure 13–3 Relationship of cement content to percentage of fine aggregate for four hypothetical trial mixes.

146 lb per cu ft, the remaining materials for the mix can be computed as follows:

Cement (from curve)		500 lb
Water (500 × 0.50)		250 lb
	Total	750 lb
Concrete per cubic yard (146 pcf × 27 cu ft)		3940 lb
Total aggregates (3940 − 750)		3190 lb
Fine aggregate (36/100×3190)		1150 lb
Coarse aggregate (3190 − 1150)		2040 lb

There are 50 batches (10 lb of cement each) per cubic yard (500 ÷ 10) and 0.54 cu ft of concrete per batch (27/50). The yield is 5.40 cu ft per 100 lb of cement (0.54 × 100/10).

Unless care is taken to control slump and air content and unless weights are accurately determined, it may be difficult to obtain data necessary to plot a curve such as Figure 13-3. For well-graded, rounded aggregates, the curve may be much more flat.

If there is a wide difference in the costs of fine and coarse aggregates, the optimum percent of fine aggregate, for economy, may be different from that at which minimum cement content occurs. For example, if coarse aggregate costs more than fine, it may be more economical to use more of the latter, with the necessary increase in cement content.

Table 13–F

SUGGESTED TRIAL MIXES FOR *NON-AIR-ENTRAINED* CONCRETE OF
(3- to 4-in. slump, well-graded aggregate with SG = 2.65)

Water-Cement Ratio, lb per lb	Maximum Size of Aggregate, in.	Air Content (Entrapped Air), Percent	Water, lb per cu yd of Concrete	Cement, lb per cu yd of Concrete
0.40	⅜	3	385	965
	½	2.5	365	915
	¾	2	340	850
	1	1.5	325	815
	1½	1	300	750
0.45	⅜	3	385	855
	½	2.5	365	810
	¾	2	340	755
	1	1.5	325	720
	1½	1	300	665
0.50	⅜	3	385	770
	½	2.5	365	730
	¾	2	340	680
	1	1.5	325	650
	1½	1	300	600
0.55	⅜	3	385	700
	½	2.5	365	665
	¾	2	340	620
	1	1.5	325	590
	1½	1	300	545
0.60	⅜	3	385	640
	½	2.5	365	610
	¾	2	340	565
	1	1.5	325	540
	1½	1	300	500
0.65	⅜	3	385	590
	½	2.5	365	560
	¾	2	340	525
	1	1.5	325	500
	1½	1	300	460
0.70	⅜	3	385	550
	½	2.5	365	520
	¾	2	340	485
	1	1.5	325	465
	1½	1	300	430

MEDIUM CONSISTENCY

With Fine Sand— Fineness Modulus = 2.50			With Coarse Sand— Fineness Modulus = 2.90		
Fine Aggregate, Percent of Total Aggregate	Fine Aggregate, lb per cu yd of Concrete	Coarse Aggregate, lb per cu yd of Concrete	Fine Aggregate, Percent of Total Aggregate	Fine Aggregate, lb per cu yd of Concrete	Coarse Aggregate, lb per cu yd of Concrete
50	1240	1260	54	1350	1150
42	1100	1520	47	1220	1400
35	960	1800	39	1080	1680
32	910	1940	36	1020	1830
29	880	2110	33	1000	1990
51	1330	1260	56	1440	1150
44	1180	1520	48	1300	1400
37	1040	1800	41	1160	1680
34	990	1940	38	1100	1830
31	960	2110	35	1080	1990
53	1400	1260	57	1510	1150
45	1250	1520	49	1370	1400
38	1100	1800	42	1220	1680
35	1050	1940	39	1160	1830
32	1010	2110	36	1130	1990
54	1460	1260	58	1570	1150
46	1310	1520	51	1430	1400
39	1150	1800	43	1270	1680
36	1100	1940	40	1210	1830
33	1060	2110	37	1180	1990
55	1510	1260	58	1620	1150
47	1350	1520	51	1470	1400
40	1200	1800	44	1320	1680
37	1140	1940	41	1250	1830
34	1090	2110	38	1210	1990
55	1550	1260	59	1660	1150
48	1390	1520	52	1510	1400
41	1230	1800	45	1350	1680
38	1180	1940	41	1290	1830
35	1130	2110	39	1250	1990
56	1590	1260	60	1700	1150
48	1430	1520	53	1550	1400
41	1270	1800	45	1390	1680
38	1210	1940	42	1320	1830
35	1150	2110	39	1270	1990

Table 13–G
SUGGESTED TRIAL MIXES FOR AIR-ENTRAINED CONCRETE OF
(3- to 4-in. slump, well-graded aggregate with SG = 2.65)[a]

Water-Cement Ratio, lb per lb	Maximum Size of Aggregate, in.	Air Content, Percent	Water, lb per cu yd of Concrete	Cement, lb per cu yd of Concrete
0.40	3/8	7.5	340	850
	1/2	7.5	325	815
	3/4	6	300	750
	1	6	285	715
	1 1/2	5	265	665
0.45	3/8	7.5	340	755
	1/2	7.5	325	720
	3/4	6	300	665
	1	6	285	635
	1 1/2	5	265	590
0.50	3/8	7.5	340	680
	1/2	7.5	325	650
	3/4	6	300	600
	1	6	285	570
	1 1/2	5	265	530
0.55	3/8	7.5	340	620
	1/2	7.5	325	590
	3/4	6	300	545
	1	6	285	520
	1 1/2	5	265	480
0.60	3/8	7.5	340	565
	1/2	7.5	325	540
	3/4	6	300	500
	1	6	285	475
	1 1/2	5	265	440
0.65	3/8	7.5	340	525
	1/2	7.5	325	500
	3/4	6	300	460
	1	6	285	440
	1 1/2	5	265	410
0.70	3/8	7.5	340	485
	1/2	7.5	325	465
	3/4	6	300	430
	1	6	285	405
	1 1/2	5	265	380

[a] Increase or decrease water per cubic yard by 3 percent for each increase or decrease of 1 in. in slump, then calculate quantities by absolute volume method. For manufactured fine aggregate, increase percentage of fine aggregate by 3 and water by 15 lb per cu yd of concrete. For less-workable concrete, as in pavements, decrease percentage of fine aggregate by 3 and water by 8 lb per cu yd of concrete.

MEDIUM CONSISTENCY

With Fine Sand— Fineness Modulus = 2.50			With Coarse Sand— Fineness Modulus = 2.90		
Fine Aggregate, Percent of Total Aggregate	Fine Aggregate, lb per cu yd of Concrete	Coarse Aggregate, lb per cu yd of Concrete	Fine Aggregate, Percent of Total Aggregate	Fine Aggregate, lb per cu yd of Concrete	Coarse Aggregate, lb per cu yd of Concrete
50	1250	1260	54	1360	1150
41	1060	1520	46	1180	1400
35	970	1800	39	1090	1680
32	900	1940	36	1010	1830
29	870	2110	33	990	1990
51	1330	1260	56	1440	1150
43	1140	1520	47	1260	1400
37	1040	1800	41	1160	1680
33	970	1940	37	1080	1830
31	930	2110	35	1050	1990
53	1400	1260	57	1510	1150
44	1200	1520	49	1320	1400
38	1100	1800	42	1220	1680
34	1020	1940	38	1130	1830
32	980	2110	36	1100	1990
54	1450	1260	58	1560	1150
45	1250	1520	49	1370	1400
39	1140	1800	43	1260	1680
35	1060	1940	39	1170	1830
33	1030	2110	37	1150	1990
54	1490	1260	58	1600	1150
46	1290	1520	50	1410	1400
40	1180	1800	44	1300	1680
36	1100	1940	40	1210	1830
33	1060	2110	37	1180	1990
55	1530	1260	59	1640	1150
47	1330	1520	51	1450	1400
40	1210	1800	44	1330	1680
37	1130	1940	40	1240	1830
34	1090	2110	38	1210	1990
55	1560	1260	59	1670	1150
47	1360	1520	51	1480	1400
41	1240	1800	45	1360	1680
37	1160	1940	41	1270	1830
34	1110	2110	38	1230	1990

Job-Size Trial Batches

On most jobs, trial batches may be full size. These batches are often used in foundations during the early phases of a construction project. The first trial may be selected on the basis of experience or from established relationships such as those given in Tables 13-F and 13-G. These tables have been developed from experience and data from several sources. The quantities are based on concrete having a slump of 3 to 4 in., and with well-graded aggregates having a specific gravity of 2.65. For other conditions, it is necessary to recalculate the quantities in accordance with the footnotes.

When making a trial batch, enough water should be added to produce the desired slump, whether it is more or less than the estimated amount. Slump, air content, and the unit weight of the fresh concrete should then be measured. A second batch to restore the desired water-cement ratio may be necessary.

Example: Assume that the mix proportions are to be determined for the following:

Conditions. A water-cement ratio of 0.53, maximum aggregate size of 1 in., air content of $6 \pm 1\%$, and a slump of 2 to 4 in. The fine aggregate has a fineness modulus of 2.85 and a free moisture content (above saturated, surface dry) of 4.4% of SSD weight. The coarse aggregate has a free moisture content of 1%.

Weights. For these conditions, the following quantities per cubic yard may be interpolated from Table 13-G:

Cement (Type IA)	540 lb
Water	285 lb
Fine aggregate	1140 lb
Coarse aggregate	1845 lb
Total	3810 lb

The free moisture in the fine aggregate amounts to 4.4% of 1140 lb, or 50 lb. The free moisture in the coarse aggregate is approximately 18 lb (use 20 lb). The corrected weights per cubic yard of concrete are:

Cement (540+0)	540 lb
Water (285−50−20)	215 lb
Fine aggregate (1140+50)	1190 lb
Coarse aggregate (1845+20)	1865 lb
Total	3810 lb

When these quantities of materials were mixed, the consistency was such that the slump was less than 1 in. Additional water (25 lb) was added to bring the slump to 3 in. Air content was measured to be 5.3% and the unit weight was measured to be 144 pcf. The workability of the mix was noted as good. The total amount of water used was 310 lb (285+25), and the total weight of concrete produced was 3835 lb (3810+25).

Volume. Since the unit weight of the concrete was 144 pcf, the volume of concrete in this batch was 26.6 cu ft (3835 lb ÷ 144 pcf).

The principal reason for the discrepancy between actual and theoretical volumes is that the specific gravity values of the aggregates were probably different from those assumed in Tables 13-F and 13-G. Therefore, these quantities must be adjusted to produce 27 cu ft of concrete, maintaining the water-cement ratio at 0.53. Unless the corrections are very large, certain simplified assumptions can be made. One, the unit weight of the concrete remains essentially constant, and two, the amount of water required per cubic yard of concrete remains constant.

Adjustment. The adjusted water requirement is

$$\frac{27}{26.6} \times 310 = 315 \text{ lb}$$

The cement requirement is

$$\frac{315}{0.53} = 595 \text{ lb}$$

The weight of materials per cubic yard of concrete must total 144×27 $\cong 3890$ lb. The total weight of aggregates must, therefore, be $3890 - (315+595) = 2980$ lb.

For the ratio of fine to total aggregates interpolated in Table 13-G (about 38%), saturated, surface-dry weights are 1135 lb and 1845 lb, respectively.

To determine the optimum proportions for economy, additional trial batches should be made varying the proportions of fine and coarse aggregates, as discussed previously.

CHAPTER 14
TRIAL MIXES—ABSOLUTE VOLUME METHOD OF MIX DESIGN

INTRODUCTION

In Chapter 13, the proportions of a concrete mixture were determined by mixing small batches in the laboratory or large batches on a job site. Using these methods, water-cement ratio, maximum size of aggregate, air content, and slump were selected to meet specific job conditions and strength and exposure requirements. The remaining variables (cement and water contents and quantities of fine and coarse aggregate) were determined by preparing a series of mixes and selecting the most economical concrete meeting all the job requirements. The only known condition of the mix materials was the moisture condition of the aggregate, saturated, surface-dry (SSD).

ABSOLUTE-VOLUME METHOD OF PROPORTIONING

The absolute volume of any loose material is the volume of solid matter. It may be computed from the material's weight and specific gravity using the equation

$$\text{absolute volume (cubic feet)} = \frac{\text{weight of loose material (pound)}}{\text{specific gravity of the material} \times \text{unit weight of water}}$$

where the unit weight of water is 62.4 lb per cu ft.

In the absolute volume method of proportioning, the volume of freshly mixed concrete is equal to the sum of the *absolute* volume of the cement plus the *absolute* volume of the aggregates (in the moisture condition used) plus the volumes of water (exclusive of any contained in the aggregate particles) and air (both entrained and entrapped).

One method of determining absolute volume of a component material is by measuring the volume of water displaced by it. This can be accomplished by partially filling a graduated container with a given number of cubic feet of water and immersing the material in the water, increasing the volume in the container. The number of cubic feet that this volume increases is the volume of the immersed material.

This method, of course, is not practical when dealing with large amounts of aggregate and cement; in such cases, volume is calculated by using the specific gravity of the material. Specific gravity is the weight of a material divided by the weight of water it displaces. For instance, if a rock weighs 4 lb and it displaces 2 lb of water, its specific gravity is $4 \div 2 = 2$; it is twice as heavy as the water it displaces. The specific gravity of a 1 cu ft, solid block of granite weighing 165 lb is its weight divided by the weight of the volume of water it displaces. Since a cubic foot (the volume of the block) of water weighs 62.4 lb, the specific gravity of the granite block is 165 lb $\div$ 62.4 lb $= 2.65$. The block is 2.65 times as heavy as the water it displaces.

The specific gravity of cement is about 3.15 and the specific gravity of most aggregates is approximately 2.65, although limestone may have a specific gravity of 2.50 or less and traprock a specific gravity of 2.75 or greater. Specific gravity can be determined in the laboratory by weighing a sample of aggregate in air then placing the same sample in a wire basket suspended from a scale. The basket containing the aggregates is then immersed in water. The difference between the weight of the sample in air and the weight of the sample immersed in water is the weight of the water the aggregate displaces. The weight of the sample in air is then divided by this weight of displaced water to obtain specific gravity. Algebraically:

$$SG = \frac{Wa}{Wa - Ww}$$

where

$$SG = \text{specific gravity}$$
$$Wa = \text{weight of sample in air}$$
$$Ww = \text{weight of sample in water}$$

When the specific gravities of the materials are determined from tests or previous known data, the absolute volume can be calculated using the formula given earlier. The specific gravity of aggregate, as used in these calculations, is the bulk specific gravity on the basis of saturated, surface-dry material (ASTM C127 and C128).

Another value commonly used is the dry-rodded unit weight of the coarse aggregate. This is a standard method (ASTM C29) for obtaining a measure of the weight of loose, dry aggregates. The procedure specifies that the aggregates be compacted into a cylindrical container of a specified size by filling the container in three layers, rodding each layer 25 times, striking off any excess aggregate level with the top of the container, and weighing the aggregate. The specified size of the container varies with the maximum size of the aggregate.

A third value sometimes useful in calculating proportions by absolute volume is the fineness modulus of the fine aggregate. The fineness modulus is an index number roughly proportional to the average size of particles in a given aggregate; that is, the coarser the aggregate, the greater the fineness modulus. It is computed by adding the cumulative percentages coarser than U.S. standard sieves Nos. 4, 8, 16, 30, 50, and 100, and dividing the sum by 100. Although the fineness modulus gives no idea of gradation and does not distinguish between a single size aggregate and a graded one having the same average size, it does indicate whether one graded aggregate sample is finer or coarser than another.

SELECTING PROPORTIONS

Regardless of the method of mix design, concrete should be placed using the minimum quantity of mixing water consistent with proper handling. This tends to improve its strength, durability, and other desirable properties. Proportions should be selected to produce concrete (1) of the stiffest consistency (lowest slump) that can be placed efficiently to provide a homogeneous mass, (2) with the maximum size of aggregate economically available and consistent with satisfactory placement, (3) of adequate durability to withstand weather and other destructive agencies to which it may be exposed, and (4) of the strength required to withstand imposed loads without danger of failure.

Slump and Maximum Size of Aggregate

The slump test is generally used as a measure of the consistency of concrete. It should not be used to compare mixes with wholly different proportions or mixes with different kinds or sizes of aggregates. When used with different batches of the same mixture, changes in slump indicate changes in materials, mix proportions, or water content. To avoid mixtures that are too stiff or too wet, slumps within limits given in Table 13-C (Chapter 13, p. 223) are suggested. Overwet mixes should always be avoided since they are difficult to place without segregation and generally result in concrete of poor durability.

Within limits of economy, the largest maximum size of aggregate should be used. The use of a larger size permits reductions in both water and cement requirements. The maximum size of coarse aggregate that can be used depends on the sizes and shapes of the concrete members to be cast, and on the amounts and distribution of reinforcing steel. The maximum size should not exceed one-fifth the minimum dimension of the member, or three-quarters the clear space between reinforcing bars or between reinforcement and forms. For unreinforced slabs on ground, the maximum size should not exceed one-third the slab thickness. Smaller sizes may be used when availability or considerations of economy require them.

Estimating Water Requirements

The amount of mixing water required to produce a cubic yard of concrete of a given slump is dependent on the maximum size of aggregate—the smaller the maximum size, the greater the amount of water required. It is therefore advisable to use the largest practical size of coarse aggregate. This minimizes the water requirement and allows the cement content to be reduced—perhaps to the minimum specified. The maximum size of coarse aggregate that produces concrete of greatest strength for a moderate cement content is about $3/4$ in. However, this depends upon aggregate source as well as its shape and gradation.

The quantities of water given in Table 13-D (Chapter 13, p. 224) apply with sufficient accuracy for preliminary estimates of proportions. They are maximums that can be expected for fairly well-shaped angular aggregates graded within the limits of accepted specifications. If other aggregates that increase mixing water requirements are used, the cement content should be increased to maintain the desired water-cement ratio.

Entrained Air

Entrained air should be used in all concrete exposed to freezing and thawing and may be used under mild exposure conditions to improve workability. It is recommended for all paving concrete regardless of climatic conditions.

The recommended total air contents for air-entrained concretes are shown in Table 13-D. Note that the amount of air required to provide adequate freeze-thaw resistance is dependent on the maximum size of the aggregate. Entrained air is contained only in the mortar portion of concrete, and in properly proportioned mixes, mortar content decreases as maximum aggregate size increases.

When mixing water is held constant, air entrainment will increase slump. When cement content and slump are held constant, less mixing water is required; the resulting decrease in the water-cement ratio helps to offset possible strength losses and results in improvements in permeability and other paste properties. Therefore, the strength of air-entrained concrete may equal, or nearly equal, that of non-air-entrained concrete when their cement contents and slumps are the same.

Minimum Cement Content

Many specifications require a minimum cement content in the concrete mix. This provides protection against the effects of increased water demands due to fluctuations in temperature or in fine aggregate gradation and thereby ensures against low strengths. Minimum cement requirements also serve to guarantee satisfactory finishability of slabs and suitable quality of vertical surfaces, as well as desired workability and strength.

The minimum cement content may be influenced by characteristics of local materials, by placement conditions, and by subsequent exposure. Where there will be little inspection, specification of a minimum cement content is recommended. If minimum strengths are also required, they should be compatible with the minimum cement content.

Selecting Water-Cement Ratio

The requirements for quality concrete may be stated in terms of durability and minimum strength, as well as in terms of minimum cement content.

While concrete durability is influenced by many variables, including mixing, placing, curing, and quality of ingredients, proportions must be selected that will ensure cement paste of adequate quality to withstand expected exposures. Suitable control of these other factors then ensures durable concrete.

Entrained air is of great benefit in ensuring durable concrete and should always be used when the concrete is exposed to severe weathering. Where concrete will be exposed to sulfate action, sulfate-resisting cement should be used (preferably Type V or, where Type V is unavailable, Type II). The critical requirement is a maximum of 5% C_3A for severe sulfate exposure; 8% for moderate. Type I cements can sometimes meet these requirements. Table 13-B (Chapter 13, p. 221) serves as a guide in selecting maximum permissible water-cement ratios for different exposures when proper use of air entrainment is made and the materials have been carefully selected.

The maximum water-cement ratio or minimum cement content needed to produce the required strength can best be determined by laboratory tests made with the same materials, including cement, to be used in the actual work. However, if it is not practical to conduct such detailed tests, Figure 13-2 and Table 13-D (Chapter 13, pp. 224 and 222) afford a basis for estimating cement and water requirements.

The required cement factor can be calculated using the maximum permissible water-cement ratio selected from Table 13-B or Figure 13-2 and water requirements from Table 13-D and dividing the pounds of mixing water required per cubic yard by the water-cement ratio. If a minimum cement content is specified, the corresponding water-cement ratio for estimating strength can be computed by dividing the pounds of water per cubic yard by the cement content in pounds per cubic yard. Selection of proportions of concrete should be based on whichever of the limitations specified—durability, strength, or cement factor—requires the lowest water-cement ratio.

Estimating the Quantity of Coarse Aggregate

The minimum amount of mixing water and the maximum strength will result for given aggregates when the largest quantity of coarse aggregate is used, consistent with adequate ease of placement and workability. This quantity can be determined most effectively from laboratory investigations of materials, subject to later adjustment in the field. However, in the absence of such laboratory data, a good estimate of the best proportions can be made for aggregates graded within conventional limits from established empirical relationships, as shown in Table 13-A (Chapter 13, p. 220). Values shown are dry-rodded bulk volumes of coarse aggregate per unit volume of concrete.

Concrete of comparable workability can be expected with aggregates of comparable size, shape, and gradation when a given dry-rodded volume of coarse aggregate per unit volume of concrete is used. In the case of different types of aggregates, particularly those with different particle shapes, the use of a fixed, dry-rodded volume of coarse aggregate automatically makes allowance for differences in mortar requirements as reflected by void content of coarse aggregate. For example, angular aggregates have a higher void content, and require more mortar than rounded aggregates. The procedure does not reflect variations in gradation of coarse aggregates within different maximum size limits, except as they are reflected in percentage of voids. The optimum dry-rodded volume of coarse aggregate per cubic yard of concrete depends on the maximum size of aggregate and the fineness modulus of fine aggregate as indicated in Table 13-A.

ABSOLUTE-VOLUME METHOD—EXAMPLE

As a means of presenting the detailed procedure of the absolute-volume method, the remainder of this chapter is devoted to an example mix-design problem. This example is worked step-by-step on Chapter 14 Data Sheets 1-10, beginning at the end of the chapter. The discussion below follows these data sheets in order.

Project-Design Data

Concrete is required for a bridge pier that will be exposed to fresh water in Buffalo, New York. The engineer's specifications require a minimum structural strength of 3000 psi at 28 days. The contract drawings indicate that the pier is to be heavily reinforced, and that 2 in. of concrete are required to cover the reinforcing steel. Type I cement is to be used for the concrete, and high-frequency vibrators have been specified for consolidation.

The technician enters this information on Data Sheet 1 of the absolute volume mix design data sheets. Then, through the use of the charts and tables which were explained in Chapter 13, he determines the basic mix design data as follows:

Water-cement ratio. By referring to Table 13-B, he notes that the type of structure, a bridge pier, falls into the category of "all other structures." Since Buffalo, New York, is located in a severe exposure area, and the concrete is to be exposed to fresh water, the required water-cement ratio for durability is 0.50 lb of water per pound of cement. The structural requirements call for a strength of 3000 psi at 28 days. Referring to the strength-band curves (Fig. 13-2), the required maximum water-cement ratio for strength (air-entrained concrete, Type I cement) is 0.57. As with the unit-weight method, the lowest water-cement ratio governs the design, and for this example the technician uses 0.50.

Aggregate size. From the information given, the maximum size of coarse aggregate should not exceed 1½ in., which is three-fourths of the distance between the reinforcing steel and the forms. If this size aggregate is not economically available, the next smaller size should be specified.

Air content. Because of the exposure conditions, the concrete should be air-entrained. The required air content, based on the maximum size of the aggregate, is obtained from Table 13-D. In this case, for 1½-in. aggregate, the air content should average 4.5% This amount of air entrainment can be achieved using either an air-entraining admixture or Type IA cement. Follow the admixture manufacturer's recommendations as to dosage. Often, dosage must be varied a bit to attain the desired level of entrained air.

Slump. The slump ranges for various types of construction are detailed in Table 13-C. Since the pier is heavily reinforced, a slump range of ¼ in. is called for. The mix will be designed for a slump of 2½ ± ½ in.

Mixing water. The only value remaining to be determined on the first mix design data sheet is the quantity of mixing water. Knowing the maximum size of the aggregate and the slump range, the technician can obtain the weight of water per cubic yard of concrete from Table 13-D. In this example, he would use about 260 lb of water per cu yd, interpolating for a 2 to 3-in. slump.

Aggregate data. With this method, it is necessary to know various characteristics of the aggregate: bulk specific gravity of fine and coarse aggregates, fineness modulus of the fines, dry-rodded unit weight of the coarse aggregate, and absorption values, and moisture values if the aggregate is wet, of fine and coarse aggregates. For this problem, these values have been previously determined and are entered at lines 1-10 on Data Sheet 2. The unit weight of the SSD coarse aggregate is calculated at line 11 and equals 110 lb per cu ft.

The absolute volume method assumes a volume of coarse aggregate based on information in Table 13-A. For this example—where the fineness modulus of the fine aggregate is 2.80, and the maximum size of the coarse aggregate is 1½ in.—the mix requires 0.71 cu yd, or 19.2 cu ft, of coarse aggregate per cubic yard of concrete, as shown in line 12. Based on the formula, weight=volume×unit weight, the weight of coarse aggregate (line 13) = 19.2×110 = 2112 lb per cu yd.

Cement content. Knowing the required amount of mixing water (260 lb per cu yd) and the water-cement ratio (0.50), cement content is determined at line 14 on data sheet 2 as follows:

$$\frac{260 \text{ lb water per cu yd}}{0.50 \text{ lb water per lb cement}} = 520 \text{ lb cement per cu yd}$$

Determination of absolute volumes. For this trial mix, it has been determined that 520 lb of cement, 260 lb of water, 2112 lb of coarse aggregate, and 4.5% air (0.045 cu yd) are required per cubic yard of concrete. These weights, together with the specific gravities of the materials, correspond to the absolute volumes shown on data sheet 3. These volumes are calculated according to the relationship, absolute volume=weight of loose material (specific gravity×62.4) and are shown at lines 15-17. When the volume of air is included (line 18), volumes total 20.71 cu ft. Weights total 2892 lb per cu yd of concrete.

The only unknowns at this point are the weight and absolute volume of fine aggregate. Since each cu. yd. of concrete contains 27 cu ft, the absolute volume of fine aggregate is equal to 27.00 minus 20.71 (total of all other materials), or 6.29 cu ft, as shown at line 20. This value, together with the specific gravity of the fine aggregate, gives a weight of fine aggregate equal to 1048 lb per cu yd of concrete (6.29×2.67×62.4). The total batch weight is, therefore, 3940 lb (2892+1048).

Since 1.22 cu ft is the volume occupied by entrained air (4.5% of 27 cu ft), the solid volume of the concrete is 27.00 minus 1.22, or 25.78 cu ft per cubic yard of concrete. This is entered at line 23 and completes Data Sheet 3.

Aggregate moisture corrections. These volumes are based on aggregate in a saturated surface-dry condition, and therefore, the mix must be adjusted for aggregate moisture of absorption. This is accomplished on Data Sheet 4. For the example problem, the aggregate is assumed to be in a wet condition. By multiplying the SSD weights of fine and coarse aggregates by their respective free moisture content percentages, the weights of surface moisture in pound per cubic yard of concrete can be obtained (lines 24a, 25a). To adjust the mix design, these amounts must be added to the SSD aggregate weights and the total weight of this surface water subtracted from the mix water weight.

Adjustment of mix design. With all of the basic data determined, it is necessary to make the mix design adjustments for moisture, as shown at lines 26-29 on Data Sheet 5. The sand percentage (using SSD weights) is calculated at line 31. Knowing the total weight of a cu yd of the concrete, unit weight can be calculated as $3940 \div 27 = 146$ lb per cu ft (line 33).

First Trial Batch

Having the design weights for 1 cu yd, it is now necessary to confirm the data by mixing a trial batch of concrete. For this example, we will assume that a quantity of concrete equal to $\frac{1}{100}$ cu yd is mixed and record the data on the lower half of Data Sheet 5. Weights of materials equal to $\frac{1}{100}$ of the values at lines 26-29 are batched; these batch weights are recorded at lines 34-37. The amount of air-entraining admixture is calculated according to the manufacturer's recommendation (lines 32, 40), and the materials are added and mixed. Only a part of the mixing water is added to the mixer, with the remaining quantity added in small amounts until the required slump is reached.

For this problem, we have assumed that the amount of mix water used is not sufficient to produce the required slump of $2\frac{1}{2} \pm \frac{1}{2}$ in. By adding 0.25 lb of additional water, a slump of 2 in. is obtained; this value is entered at line 41 and the appropriate note is referenced (Note b, line 46). Measured unit weight is 145 lb per cu ft (line 42), and measured air content, 5.2% (line 43). We record the mix as having a fair workability and rocky appearance (lines 44, 45).

It is now necessary to adjust the trial-mix design. If no water remained to be added and slump was not excessive, or if additional water was not needed to meet the required slump, the mix design would have been confirmed.

Absolute Volume Mix Design Adjustments

Water and Cement. Initial completion of the trial mix showed that 0.25 lb of additional water are required to achieve the necessary slump. In order to maintain a constant water-cement ratio of 0.50, the cement content must also be increased.

On Data Sheet 6, water content is adjusted by multiplying the amount of additional water needed (0.25 lb) by the number of batches per cubic yard (100). The resultant figure (25 lb) is the additional amount of water to be added per cubic yard (line 49).

The new cement content is determined by adding the original design weight of water (260 lb) and the adjustment weight (25 lb) and dividing the sum (285 lb) by the water-cement ratio (0.50). The adjusted cement content is then 570 lb per cu yd of concrete, as shown at line 54.

Determination of Absolute Volumes

On Data Sheet 7, absolute volumes are recomputed using the adjusted cement and water weights. The procedure is identical to that for original absolute volume computation on Data Sheet 3. A subsequent adjustment in air-entraining agent dosage will be made following the method outlined in the note and line 64.

Aggregate Moisture Corrections

Because the original absolute volumes were based on SSD aggregate condition, moisture adjustment (as indicated on Data Sheet 8) is necessary. The calculations for surface moisture weights are similar to those on Data Sheet 4. The adjusted figures are carried over to Data Sheet 9 where, at lines 67-70, the mix design is adjusted as it was on Data Sheet 5. Percentage of fine aggregate (using SSD weights) is calculated at line 72 and unit weight (145 lb per cu ft) at line 74.

Second Trial Batch

The second mix, like the original trial batch, should equal $\frac{1}{100}$ cu yd, and individual batch weights must be redetermined. Batch weights are shown on Data Sheet 9, lines 75-78. These amounts of cement, fine and coarse aggregates, and water can be batched and the concrete mixed with the adjusted amount of air-entraining admixture (1.7 ml) from line 64, Data Sheet 7. After this the necessary tests can be performed. The results are recorded at lines 82-86 on Data Sheet 9. For this trial mix it is assumed that slump was measured as $2\frac{1}{4}$ in.; unit weight, 145 lb per cu ft; and air content, 4.7%. Workability and appearance are both recorded as good. As can be seen, the mix design is confirmed.

Once the mix design is confirmed, two additional calculations are necessary: yield and gravimetric air content.

Yield

Yield (cubic foot concrete per 100 lb cement) is a function of measured unit weight (pound per cubic foot), total batch weight (pound), and cement content (pound). This mix, as shown at lines 87 and 88 on Data Sheet 10, produces 4.72 cu ft of concrete per 100 lb (1 cwt) of cement.

Gravimetric Air Content

Gravimetric air content, determined as a function of air-entrained and non-air-entrained (solid volume) unit weights, provides a general check in the field during construction if air measurements cannot be made. Knowing the gravimetric air content, it is only necessary to determine the unit weight of the concrete to obtain an approximation of the air content in it. For this example, gravimetric air content is 3.97%, as calculated at lines 89 and 90 on Data Sheet 10.

Summary and Strength Prediction

The final results of the mix are confirmed after cylinders are tested. For this example, the 7-day compressive strength is assumed to be 2550 psi. The 7-day strength is generally considered to be two-thirds of the 28-day strength; therefore, the anticipated 28-day strength can be determined by multiplying the 7-day strength (2550 psi) by a factor of 1.5 to obtain a strength of 3830 psi, well in excess of the required 3000 psi. The 7-day compressive strength is entered at the bottom of Data Sheet 10 along with the final record of the confirmed mix design including maximum aggregate size, percentage fine aggregate, slump, and yield. If the concrete fails to meet required strength, there could be many sources of error that the technologist would be required to look for and correct before proceeding with full-scale batches on the job. For this reason it is highly recommended that mixes be designed well in advance of actual construction.

Data Sheet 1

CONCRETE MIX DESIGN—ABSOLUTE VOLUME METHOD

Job Location: _Buffalo, N.Y._

Concrete to be Used for _Bridge pier--fresh water crossing_

Sketch of Concrete Item:

1. _Bridge pier to be heavily reinforced_
2. _Type I cement_
3. _High-frequency vibrators required_

Type of Cement: _I_

Exposure Conditions:

Type of Structure or Use: _Moderate section_

Severe Exposure _✓_ Mild Exposure ____

In Air ____ In Fresh Water _✓_ In Sea Water or in Contact with Sulfates ____

Maximum W/C for Exposure: ____ _0.50_ ____ lb water/lb cement

Specified Strength Design: _3,000_ psi at _28_ days

W/C for Strength: _0.57 (A/E)_ lb water/lb cement

W/C to Use: ____ _0.50_ ____ lb water/lb cement

Maximum Aggregate Size: _0.75 × 2" = 1½"_

Air Content: Entrained _4.5_ % Non-Air-Entrained ____%

Recommended Slump: Min. _1_ in. Max. _4_ in. Use _2½_ in. _± ½"_

Mixing Water: ____ _260_ ____ lb cu yd of concrete

Data Sheet 2

AGGREGATE DATA

(1) Bulk Specific Gravity, SSD Basis, of Fine Aggregate: $\underline{2.67}$ (SG)

(2) Weight of 1 cu ft (Solid Volume) of Fine Agg.
Calculate: SG×62.4=$\underline{2.67}$×62.4=$\underline{166.5}$ lb/cu ft

(3) Bulk Specific Gravity, SSD basis, of Coarse Agg.: $\underline{2.67}$ (SG)

(4) Weight of 1 cu ft (Solid Volume) of Coarse Agg.
Calculate: SG×62.4=$\underline{2.67}$×62.4=$\underline{166.5}$ lb/cu ft

(5) Fineness Modulus of Fine Aggregate: $\underline{2.80}$

(6) Dry-Rodded Unit Weight of Coarse Aggregate: $\underline{108}$ lb/cu ft

(7) Coarse Aggregate Absorption: $\underline{2.0}$ %

(8) Fine Aggregate Absorption: $\underline{2.0}$ %

If Aggregates are Wet, Determine the Following:
(9) Free Moisture* Content of Fine Aggregate: $\underline{4.0}$ %

(10) Free Moisture* Content of Coarse Aggregate: $\underline{1.0}$ %

*Percentage of Moisture above SSD=
$$\frac{\text{Wt. of Wet. Agg.}-\text{Wt. of SSD Agg.}}{\text{Wt. of SSD Agg.}} \times 100$$
Unit Weight of SSD Coarse Aggregate=
Dry-Rodded Unit Wt. (from line 6) × (1+% Coarse Agg. Absorption/100) =

(11) $\underline{108}$ (1+ $\frac{\underline{2}}{100}$) = $\underline{110}$ lb/cu ft

Coarse Aggregate for 1 Cu Yd of Concrete:

(12) Volume of Coarse Aggregate: $\underline{19.2}$ cu ft (rodded)/cu yd of concrete
Weight of Coarse Aggregate=Volume of CA×Unit Weight SSD=

(13) $\underline{19.2}$ × $\underline{110}$ = $\underline{2,112}$ lb/cu yd of concrete

Cement Content= $\dfrac{\text{Mixing Water (lb/cu yd of concrete)}}{\text{W/C (lb water/lb cement)}} = \dfrac{\underline{260}}{\underline{0.50}} =$

(14) $\underline{520}$ lb/cu yd of concrete

Data Sheet 3

DETERMINATION OF ABSOLUTE VOLUMES

$$\text{Absolute Volume} = \frac{\text{Weight of loose material}}{\text{SG} \times 62.4}$$

(15) Cement: $\underline{520}$ lb/cu yd of concrete ÷ ($\underline{3.15} \times 62.4$) = $\underline{2.64}$ cu ft

(16) Water: $\underline{260}$ lb/cu yd of concrete ÷ (1.00×62.4) = $\underline{4.17}$ cu ft

(17) CA: $\underline{2,112}$ lb/cu yd of concrete ÷ ($\underline{2.67} \times 62.4$) = $\underline{12.68}$ cu ft
(SSD, from line 13)

(18) Air Content: $\underline{4.5}$ % × 27 cu ft ÷ 100 = $\underline{1.22}$ cu ft

(19) Subtotals: Weight = $\underline{2,892}$ lb/cu yd of concrete; Volume = $\underline{20.71}$ cu ft

(20) FA (SSD): Volume = 27.00 − $\underline{20.71}$ = $\underline{6.29}$ cu ft
Weight = Volume × SG × 62.4 =

(21) $\underline{6.29}$ cu ft × $\underline{2.67}$ × 62.4 = $\underline{1,048}$ lb/cu yd of concrete

(22) Total Batch Weight = $\underline{3,940}$ lb/cu yd   Total Abs Vol = $\underline{27.00}$ cu ft
Concrete Solid Volume = Concrete Vol (cu ft) − Air Vol. (cu ft)

$$= 27.00 − \underline{1.22} \qquad\qquad = \underline{25.78} \text{ cu ft}$$

The above weights are for aggregates in the saturated, surface-dry condition. When the aggregates for the trial batch must be used in either the wet or dry condition, the following adjustments (sheets 4 and 5) will convert SSD weights to equivalent weights of the aggregate in a wet or dry condition.

Data Sheet 4

AGGREGATE MOISTURE AND ABSORPTION CORRECTIONS

DRY Fine or Coarse Aggregate absorbs mix water to attain the SSD condition. The weight of this absorbed water is approximately equal to

$$\text{SSD weight} \times \frac{\% \text{ Absorption}}{100} = \text{Weight Water Absorbed}$$

	SSD Wt. (lb/cu yd of concrete)	% Absorption/ 100	Wt Water Absorbed (lb/cu yd of concrete)
(24) Fine Aggregate	_____ ×	_____ 100	= _____
(25) Coarse Aggregate	_____ ×	_____ 100	= _____

To adjust batch weights for DRY aggregate, *subtract* the weight of absorbed water from the SSD aggregate weight. *Add* the total absorbed weight of water to the calculated design mix water weight on sheet 5.

With *WET Fine or Coarse Aggregate*, the surface moisture enhances the water content of the mix. Its weight is calculated by the following equation:

$$\text{SSD weight} \times \frac{\% \text{ Moist. Content}}{100} = \text{Weight Surface Moisture}$$

	SSD Wt. (lb/cu yd of concrete)	% Moist. Content/100	Wt Surface Moisture (lb/cu yd of concrete)
(24a) Fine Aggregate	*1,048* ×	*4* 100	= *41.9*
(25a) Coarse Aggregate	*2,112* ×	*1* 100	= *21.1*

To adjust batch weights for WET aggregate, *add* the weight of surface moisture to the SSD weight and *subtract* the total weight of surface moisture from the calculated design mix water weight on sheet 5.

For *SSD Fine and Coarse Aggregates*, make no water adjustments. Use the design weights.

Data Sheet 5

ADJUSTMENT OF MIX DESIGN BEFORE TRIAL BATCH
FOR SURFACE MOISTURE OR WATER ABSORPTION OF AGGREGATES

Weights (lb) per Cubic Yard of Concrete

	Design Wt. for SSD Agg	Moist. or Abs. Corrections	Adjusted Batch Weights
(26) Cement:	520		520
(27) Fine Aggregate:	1,048 (27a)	+ or – 42	= 1,090
(28) Coarse Aggregate:	2,112 (28a)	+ or – 21	= 2,133
(29) Water:	260	– or + 63	= 197

(Added Water)

(30) TOTAL 3,940

(31) Percentage FA $= \dfrac{(27a)}{(27a)+(28a)} = \dfrac{1,048}{1,048 + 2,112} = \underline{33.2}$ %

(32) Air-Entraining Admixture manufacturer's recommended dosage: $6\frac{1}{2}$ oz/cu yd of concrete

$$\text{Calculated Unit Weight} = \frac{\text{Total Wt (lb/cu yd of concrete)}}{27 \text{ (cu ft/cu yd)}} =$$

(33) $\dfrac{3,940}{27} = \underline{146}$ lb/cu ft

Trial Mix Batch Weights based on 1/100 cu yd ✓; 1/25 cu yd ___; 1 cu ft ___

(34)	Cement:	5.20 lb
(35)	Fine Aggregate:	10.90 lb
(36)	Coarse Aggregate:	21.33 lb
(37)	Water:	1.97 lb
(38)	Total Wt per batch:	39.40 lb
(39)	Number of batches per cubic yard:	100

AE Admixture (ml/batch) = AE Admixture (oz/cu yd of concrete) × $= \dfrac{30}{\text{No. batches}}$

Data Sheet 5 (Continued)

(40) $\dfrac{6\frac{1}{2} \times 30}{100} = \underline{\quad 2 \quad}$ ml/batch

Mix Information

(41)   Measured slump: _2.0_ in. See Note _6_, below

(42)   Measured unit weight: _145_ lb/cu ft

(43)   Measured air content _5.2_ %

(44)   Workability: Good _____ Fair _✓_ Poor _____

(45)   Appearance: Oversanded _____ Good _____ Rocky _✓_

Notes

(a) If the slump and air content are reached after the calculated weight of water has been added to the mix, the mix design has been confirmed. Proceed to Sheet 10 and complete the data and cast test cylinders.

(46)   (b) _0.25_ lb of additional water are needed to reach the required slump. Proceed to Sheet 6 and adjust the mix.

(47)   (c) _____ lb of water are left over after the slump is reached. Proceed to Sheet 6 and adjust the mix.

ADJUSTMENT OF MIX DESIGN AFTER TRIAL BATCH

Additional Water Needed to Reach Required Slump
Increase the cement content to maintain the water-cement ratio.
Actual water used in the mix:

(48)   Design Weight Water: *260* lb (from line 16)
    Plus No. of batches per cu yd (line 39) × Added water wt (line 46) =

(49) $\underline{\quad 100 \quad} \times \underline{\quad 0.25 \quad} = +$ *25* lb

(50)   Total Water in New Mix: *285* lb/cu yd of concrete

Water is Left Over After Reaching Slump
Reduce the cement content to maintain the water-cement ratio.
Actual water used in the mix:

(51)   Design Weight Water: _____ lb (from line 16)
    Minus No. of batches per cu yd (line 39) × withheld water weight (line 47) =

(52) $\underline{\qquad} \times \underline{\qquad} = -$ _____ lb

(53)   Total Water in New Mix: _____ lb/cu yd of concrete

Adjustment of Cement Content
    Adjust the mix using the new total water value (line 50 or 53), and the required W/C to obtain a new cement content.

$$\text{New Cement Content} = \frac{\text{Total Water in New Mix (lb/cu yd of concrete)}}{\text{Required W/C (lb water/lb cement)}}$$

(54) $= \dfrac{285}{0.50} = \underline{\quad 570 \quad}$ lb/cu yd of concrete

DETERMINATION OF ABSOLUTE VOLUMES

(55) Cement: _570_ lb/cu yd of concrete ÷ ($\underline{3.15}$ × 62.4) = $\underline{2.90}$ cu ft

(56) Water: _285_ lb/cu yd of concrete ÷ (100 × 62.4) = $\underline{4.57}$ cu ft

(57) CA: $\underline{2,112}$ lb/cu yd of concrete ÷ ($\underline{2.67}$ × 62.4) = $\underline{12.68}$ cu ft
(SSD, from line 13)

(58) *Air Content: $\underline{4.5}$ % × 27 ÷ 100 = $\underline{1.22}$ cu ft

(59) Subtotals: Weight = $\underline{2967}$ lb/cu yd of concrete; Volume = $\underline{21.37}$ cu ft

(60) FA (SSD): Volume = 27.00 − $\underline{21.37}$ = $\underline{5.63}$ cu ft
Weight = Volume × SG × 62.4 =

(61) $\underline{5.63}$ cu ft × $\underline{2.67}$ × 62.4 = $\underline{938}$ lb/cu yd of concrete

(62) Total Batch Weight = $\underline{3.905}$ lb/cu yd   Total Abs Vol = $\underline{27.00}$ cu ft
Concrete Solid Volume = Concrete Vol (cu ft) − Air Vol (cu ft)

(63)   = 27.00 − $\underline{1.22}$ = $\underline{25.78}$ cu ft

*Note:

If concrete is not air entrained, enter the percentage of air found in the last trial concrete mix. If the concrete is air entrained and the air content found with the last trial mix differs from the desired value by not more than ½%, enter that exact air content percentage and make no change in the dosage of the air-entraining agent. If the trial-mix air content is considerably different from that desired for air-entrained concrete, leave the percentage of air as it was noted initially in line 18, and change the dosage of air-entraining agent according to the following equation:

$$\text{New AE agent dosage (ml)} = \frac{\text{Desired \% Air}}{\text{Trial Mix \% Air}} \times \text{Initial AE agent dosage (ml)}$$

(64) $= \dfrac{\underline{4.5}}{\underline{5.2}} \times \underline{2} = \underline{1.7}$ ml

AGGREGATE MOISTURE AND ABSORPTION CORRECTIONS

DRY Fine or Coarse Aggregate absorbs mix water to attain the SSD condition. The weight of this absorbed water is approximately equal to

$$\text{SSD weight} \times \frac{\% \text{ Absorption}}{100} = \text{Weight Water Absorbed}$$

	SSD Wt. (lb/cu yd of concrete)	% Absorption/ 100	Wt Water Absorbed (lb/cu yd of concrete)
(65) Fine Aggregate	_____	× _____ / 100	= _____
(66) Coarse Aggregate	_____	× _____ / 100	= _____

To adjust batch weights for DRY aggregate, *subtract* the weight of absorbed water from the SSD aggregate weight. *Add* the total absorbed weight of water to the calculated design mix water weight on sheet 5.

With *WET Fine or Coarse Aggregate,* the surface moisture enhances the water content of the mix. Its weight is calculated by the following equation:

$$\text{SSD weight} \times \frac{\% \text{ Moist. Content}}{100} = \text{Weight Surface Moisture}$$

	SSD Wt. (lb/cu yd of concrete)	% Moist. Content/100	Wt Surface Moisture (lb/cu yd of concrete)
(65a) Fine Aggregate	*938*	× *4* / 100	= *37.5*
(66a) Coarse Aggregate	*2,112*	× *1* / 100	= *21.1*

To adjust batch weights for WET aggregate, *add* the weight of surface moisture to the SSD weight and *subtract* the total weight of surface moisture from the calculated design mix water weight on sheet 9.

For *SSD Fine and Coarse Aggregates,* make no water adjustments. Use the design weights.

ADJUSTMENT OF MIX DESIGN BEFORE TRIAL BATCH
FOR SURFACE MOISTURE OR WATER ABSORPTION OF AGGREGATES

Weights (lb) per Cubic Yard of Concrete

	Design Wt for SSD Agg	Moist. or Abs. Corrections	Adjusted Batch Weights
(67) Cement:	570		570
(68) Fine Aggregate:	938 (68a) + or – 37	=	975
(69) Coarse Aggregate:	2,112 (69a) + or – 21	=	2,133
(70) Water:	285 – or + 58	=	227

(Added Water)

(71)   TOTAL   3,905

(72) Percentage FA $= \dfrac{(68a)}{(68a)+(69a)} = \dfrac{938}{938 + 2,112} = $ _____ 30.7 _____ %

(73) Air-Entraining Admixture manufacturer's recommended dosage: 6½ oz/cu yd of concrete

Calculated Unit Weight $= \dfrac{\text{Total Wt (lb/cu yd of concrete)}}{27 \text{ (cu ft/cu yd)}} =$

(74) $\dfrac{3,905}{27} = 145$ lb/cu ft

Trial Mix Batch Weights based on 1/100 cu yd ✓ ; 1/25 cu yd _____ ; 1 cu ft _____

(75)   Cement: 5.70 lb

(76)   Fine Aggregate: 9.75 lb

(77)   Coarse Aggregate: 21.33 lb

(78)   Water: 2.27 lb

(79)   Total Wt per batch: 39.05 lb

(80)   Number of batches per cubic yard: 100

AE Admixture (ml/batch) =

AE Admixture (oz/cu yd of concrete) $\times \dfrac{30}{\text{No. batches}} =$

(81)

$= $ _____ 1.7 _____ ml/batch

Mix Information

Data Sheet 9 (Continued)

(82)   Measured slump: _2¼_ in.

(83)   Measured unit weight: _145_ lb/cu ft

(84)   Measured air content: _4.7_ %

(85)   Workability: Good _✓_ Fair _____ Poor _____

(86)   Appearance: Oversanded _____ Good _✓_ Rocky _____

> *Note:* Mix design should be confirmed after all the water has been added. If desired slump is not obtained, report whatever slump has been reached, and cast test cylinders.

Data Sheet 10

DATA FROM FINAL ADJUSTED BATCH (Sheet 5 or 9)
Yield (ASTM): Volume (cu ft) of concrete produced per 100 lb of cement

$$\text{Batch Volume} = \frac{\text{Total wt of batch}}{\text{Measured Unit Wt (from line 42 or line 83)}} =$$

(87) $\dfrac{3,905}{145}$ = 0.269 cu ft

$$\text{Yield} = \text{Batch Volume} \times \frac{100}{\text{Cement Content}} =$$

(88)

$\underline{\quad 0.269 \quad}$ $\dfrac{\times\ 100}{5.70}$ = 4.72 cu ft/cwt

Gravimetric Air Content: Unit weight of concrete without air (UWN) minus actual air-entrained concrete unit weight (UW), divided by unit weight of concrete without air (UWN). UWN calculated from values obtained from lines 22-23 on sheet 3 or from lines 62-63 on sheet 7.

(89) $\text{UWN} = \dfrac{\text{Total Batch Weight (lb)}}{\text{Concrete Solid Vol (cu ft)}} = \dfrac{3,905}{25.78} = \underline{\quad 151 \quad}$ lb/cu ft

(90) $\% \text{ Air} = \dfrac{\text{UWN}-\text{UW}}{\text{UWN}} \times 100 = \dfrac{151-145}{151} \times 100 = \underline{\quad 3.97 \quad} \%$

Final Record of Confirmed and Adjusted Mix Design Data:

W/C: 0.50 lb water/lb cement   Percentage Fine Agg: 31 %

Water Added: 227 lb/cu yd   Cement Factor: 570 lb/cu yd
                of concrete   of concrete

Max. Size of Agg: $1\frac{1}{2}$ in.   Yield: 4.72 cu ft/100 lb cement

Slump, Actual: $2\frac{1}{4}$ in.   Measured Air: 4.7 %

Compressive Strength at 7-day Cylinder Break: $2,550$ psi

CHAPTER 15
QUANTITY BATCHING
AND MIXING

INTRODUCTION

When the materials for making a concrete mix have been obtained and their proportions calculated, actual production then requires batching and mixing.

Batching consists of measuring, by weight or by volume, the quantities of materials desired in the concrete and introducing these materials into the mixer.

ON-THE-JOB BATCHING

Volumetric Batching
Until the late 1920s, planned proportioning of concrete mixes was practically nonexistent. Proportioning was strictly volumetric—seldom accurate. Shovels and wheelbarrows were the accepted units for measuring aggregates. The bag (1 cu ft or 94 lb) was the measuring unit for cement. Most commonly used proportions were 1:2:4, 1:2 to ½:5, and 1:3:6— volumes of cement, sand, and coarse aggregate respectively. Bags of cement were split by hand, and every adept mixer operator had two inept counterparts. Mixers were, for the most part, steam-driven and portable.

All concrete was prepared at a job site, discharged into wheelbarrows, and transferred directly into forms (Fig. 15-1).

Water was added to mixers before and during actual mixing. Amounts added were determined by the concrete foreman's visual analysis of consistency. He threw in pailfuls of water until the concrete "looked right." Moisture content of sand and aggregates was unheard of. Sand was sand; rocks were rocks.

In an effort to improve on these crude proportioning methods, some specification writers and government agencies required the use of measuring boxes. These wooden containers were designers to be filled and struck before being added to a mix. Because work with the measuring boxes was tedious, laborious, and time consuming, contractors avoided using them whenever possible (Fig. 15-2).

Wheelbarrow Scale

The introduction of weight batchers for large jobs by the road builder prepared the way for the use of wheelbarrow scales for proportioning of

Figure 15–1 Steam-driven mixer on a job site in 1923.

Figure 15–2 Box measuring was one of the first attempts to introduce more accuracy in batching concrete.

concrete aggregates for smaller applications. These scales were used extensively in the construction of commercial and industrial building footings, reinforced concrete framing, floor slabs, and sidewalks. Wheelbarrow scales were the most practical means available before the advent of the ready-mix plant for all concrete construction (except roads, large dams and other mass concrete structures).

They were particularly useful for the following:

1. Remote or inaccessible areas.
2. Small yardages or small pours in less remote regions where the long haul of ready mix concrete could be justified.
3. Construction in countries where there was an abundance of cheap labor.

Wheelbarrow scales were equipped with a tare beam, three weight beams (generally of 500-lb capacity each), and a balance indicator. The counterweight on the tare beam was adjusted to balance out or compensate for the weight of the wheelbarrow; the beam lifter permitted the operator to add or remove any one or a combination of the weight beams from the scale system (Fig. 15-3).

The weight of each material required in the mix was set on the weight beams, one to a beam, by adjusting the position of the poise weights. In this way, the operator could weigh sand, stone or cement without

Figure 15–3 The wheelbarrow scale was one of the first ways of accurately proportioning ingredients for smaller applications.

changing the position of any of the poise weights—by merely unlocking the proper weight beam. The balance indicator showed whether the material in the wheelbarrow exactly balanced the weight set on the weight beam in action.

The wheelbarrow scale, low in cost and used in conjunction with a small construction mixer, constituted the smallest on-the-job concrete plant capable of producing controlled, quality concrete. Production capacity was established by the mixer size.

Trolley Batcher Plants

Development of the front-end loader brought about a marked improvement in on-the-job plant operations. The loader made feasible the introduction of the trolley batcher plant—a low, steel bin equipped with a track, special scale, and weigh hopper. Bins ranged in capacity from 8 to 40 tons, with some units as small as 3-ton capacity available. Batchers were designed to carry ½ to 2 tons, and were usually equipped with two to four weigh beams for cumulative weighing.

A trolley batcher operator could successively weigh up each material, pushing the weigh hopper from bin gate to bin gate along the track. Then, pushing the weigh hopper on the same track to the hopper's extended position over a mixer skip and adding the proper bagged cement, he could complete the batch.

Portable bin and batcher plants had an important place in construction for many years. They filled the gap between wheelbarrow operations and the ready mix and permanent plant periods. Currently, their use is, for the most part, limited to precast concrete plants and remote or inaccessible areas where volume or rate of pour do not warrant conventional portable batching plant facilities.

HIGH-PRODUCTION BATCHING

Batchers are essentially the same whether they are used in road builders' plants, ready-mix plants, or mass-concrete plants; they differ only in adaptations provided to meet special requirements.

Thorough mixing of concrete is more efficient when all materials are charged into the mixer at approximately the same time. This is true regardless of the size or type of mixer because of the partial blending materials receive during charging. Before materials can be simultaneously charged into a mixer, each must be measured. This can be accomplished using either separate hoppers or one common hopper. Separate hoppers, known as single-material batchers, emphasize batching by volume. Common hoppers, also called cumulative batchers, batch by weight. Although these common hoppers are readily applicable to aggregate addition, cement is almost exclusively weighed separately.

Volumetric Batchers

Volumetric batching reached its greatest efficiency and accuracy with the introduction of a batcher equipped with an adjustable cover cone and measuring stick. This batcher was a cylindrical metal hopper with a conical bottom and an adjustable conical cover. The bottom of the hopper contained a discharge gate; the cover, which had a small opening in the top, could be raised or lowered to increase or decrease its capacity of sand or stone. A measuring stick, calibrated in cubic feet, was provided to allow setting the cover to a proper position for desired cubic content.

Volumetric batching was never satisfactory even in the early days of proportioning of concrete materials. Increased need for speed and accuracy dictated that volumetric batchers would be displaced by weigh batchers. Since sand swells or "bulks" as its moisture content increases, adjustments in the volume of the sand batcher to compensate for these changes in dampness are necessary. Means to facilitate these adjustments were provided, but the operation was never entirely satisfactory.

During the period of transition from measurement by volume to measurement by weight (volumetric to gravimetric), many batchers were produced that were merely volumetric batchers fitted with a scale lever system and that could be used either volumetrically or gravimetrically.

Inundation Batchers

Inundation batchers were used extensively on large jobs in the 1920s. They were developed to eliminate one of the greatest difficulties encountered in the use of volumetric batchers. Sand with a moisture content of 5% by weight occupies about 25% more space than this same sand does when dry. The variation in volume between dry and moist sand can vary from 15 to 40%. The inundation system used a volumetric sand batcher, partly filled with water, that sifted sand into the hopper. Since sand immersed in water occupies the same total space as dry sand in air, the inundation batcher measured sand for concrete batches with considerable accuracy. Water slightly in excess of that required to saturate the sand was charged into the sand inundator; the excess was wasted. Any additional water required for the mix was batched from a separate volumetric water batcher.

Single-Material Weigh Batchers

Single-material weigh batchers are simple weighing devices designed for aggregate. They consist of a supporting frame that holds a fill gate and a scale lever system. These support the weigh hopper equipped with a discharge gate. The entire structure is bolted directly to the bottom of a storage bin.

The size of the fill valve is important. It must be large enough to allow rapid batching without catching larger aggregates or causing sand to arch.

Weigh hoppers are generally constructed of sheet steel, either circular or rectangular in section, with a bottom sloping toward the discharge gate. The gate may be of the guillotine, radial, or clam types, or it may be a simple hinged flat plate that drops away when the latch is released. All have their merits and situations where they best meet the needed requirements. For a controlled-discharge rate, a batcher should be equipped with a clam gate.

Manual, semiautomatic, single-material batchers are equipped with fill valves that are opened by the operator but close automatically when near-final weight is achieved. The valve must be manually cranked to reach final desired weight. These devices allow an operator to work more than one batcher at a time. Air-operated valves and gates are available to prevent operator fatigue and provide remote operation.

Cement Batchers

Cement batchers require an altogether different type of valve; one that is cement tight. This is generally of the plug type, with the plug rotating on a shaft to open and close the valve. A slide-type emergency valve should be installed above the cement-plug valve to allow the flow of cement from the storage bin to be cut off. This permits removal of hard lumps of cement or foreign objects that do not pass through the restricted opening of the cement valve.

Cement weigh hoppers differ from aggregate hoppers primarily in that they must have a tight cover and a cement-tight bottom gate. A canvas sleeve is usually provided to connect the fill valve to the hopper top, preventing loss of cement when charging the hopper. Such a sleeve can also materially reduce dusting. Cement hoppers require vents to allow the escape of air when charging and the return of air when discharging. Without them, hoppers can become air bound.

Discharge gates are available in various designs. Some are of the hinged flat plate type that seal against rubber; others are metal cones sealed against the rubber on the bottom of the hopper. One manufacturer makes a simple but effective gate by attaching a flexible rubber tube to the hopper bottom. There are two movable horizontal bars or rollers below the hoppers that squeeze the tube between them to cut off the flow of cement.

Single-material batcher operation is simple because the majority of single-material batchers in use are manually operated. Good operators can avoid mistakes and can weigh materials rapidly.

Multiple or Cumulative Batchers

Cumulative batchers are designed to weigh up materials one after another in a common weigh hopper suspended from a single scale-lever system. Cumulative batching is normally restricted to aggregates only.

If the scale is equipped with a dial for direct reading of the weight, the weighing operation differs from that of a scale equipped with weigh-beams and a balance indicator. Multiple batchers with full reading dials are less complex and more efficient. If a batch requires 1400 lb of sand, 700 lb of stone No. 1, and 1000 lb of stone No. 2, and the batcher is dial-equipped, the operator first weighs up the 1400 lb of sand. Then he adds stone No. 1 until the dial pointer registers at the 2100-lb mark. He comples the batch by adding stone No. 2 until the dial reads 3100 lb.

Multiple batchers with weighbeams require balance indicators. There is always one weighbeam and one fill valve for each material. The weighbeams are connected to each other by linkages, and the top beam is connected by another similar linkage to the scale lever system. Latches (one for each weighbeam) are provided to lock out all weighbeams. For the example batch above, the poise weight on the top weighbeam would be set for 1400 lb, the poise weight on the second weighbeam for 700 lb, and the poise weight on the third for 1000 lb. The top or first beam is the sand beam, the second, the stone No. 1 beam, and the third, the stone No. 2 beam.

The operator would first unlock the sand weighbeam throwing the scale out of balance. Sand then pours into the hopper until the indicator shows the scale to be in balance, indicating that the weight of the sand in the weigh hopper equals that of the poise weight setting. Leaving the sand beam unlocked, or on the scale system, he would next unlock the stone No. 1 weighbeam, throwing the scale out of balance, and add the stone to the weigh hopper until the scale is again in balance.

The two weighbeams exert a combined effect on the scale system of 2100 lb. This weight of sand and stone No. 1 is needed to bring the scale into balance. He would complete the batch by unlocking the third weighbeam and adding stone No. 2, following the same process.

Multiple batcher sizes range from 1 to 6 cu yd or more, and they can be hand arranged for two, three, four, five, or six or more materials. The larger batchers are available on special order. Size ratings are based on the size of the concrete batch rather than the cubic content of the weigh hopper.

Batcher scale-lever systems typically have the weigh hopper suspended by four hanger rods, carried by four pivots or knife edges, two on each of the collecting levers. These collecting levers are sometimes referred to as rocker arms. The weight of the hopper and its contents, reduced in accordance with the ratio of the collecting levers, is applied to the second lever in the system. Again the load is transmitted to the next lever at a reduced value according to the ratio of the second lever.

A specific method is recommended by NRMCA for distinguishing between manual, semiautomatic, and automatic batching systems: In an automatic system, the entire sequence for measurement of all major ingredients is actuated by a single operation—pushing a button or inserting a card. After this the cycle is completed without further attention. In a semiautomatic system, the weighing of an ingredient is actuated separately by the operator but is terminated automatically when the proper quantity is reached. In manual operation, cutoff of a material at the proper quantity is accomplished by the operator. The system is also classed as manual if any major ingredient—cement, aggregate, or water—is batched manually.

Interlocks, which prevent batching gates from opening while material is being discharged from batchers and the batcher gates from opening during the weighing cycle, are used on semi- and totally automatic batchers. Provision for a visual check of each scale reading before the next weighing cycle is important. In automatic batching, the interlock prevents the charging device from being activated until the scale has returned to zero balance after a batch. Graphic or digital recorders that tabulate the weight of each material are essential where accurate records are required. They are mandatory with many state highway departments.

WEIGHING TOLERANCES

When a batch is weighed up either manually or automatically, the material weights are seldom absolutely correct. There are two major kinds of errors that make correct weighing impossible: scales are never perfect (they should be periodically calibrated) and slight errors always exist in any weighing operation.

Errors resulting from an operator's fallibility in manual weighing or from mechanical devices in the case of automatic weighing not cutting off the flow of material into the weigh hopper at the exactly correct point are much greater than any error in the scale itself.

Another consideration in batching is the weight of material "in transit" between the fill gate and its placement in the weigh hopper. In manual operations the operator allows for this, that is, he closes the gate before exact final weight is reached. If he underestimates, he can open the fill gate to allow a small amount of additional material to flow through.

Although requirements are constantly becoming more stringent, precise measurement of each ingredient of concrete is neither economically desirable nor possible.

ASTM C94 specifies tolerances for ready mixed concrete that have been universally accepted. They are taken from the National Bureau of Standards' *Handbook 44: Specification, Tolerances, and Other Technical Requirements for Commercial Weighing and Measuring Devices.* Most recommended tolerances do not exceed ±1% for cement and water, ±2% for each aggregate, and ±3% for admixtures.

Water for concrete should be measured by automatic meters or water weigh batchers. Diverter valves are needed for double checking the weights.

Liquid admixtures are usually added in small doses; therefore, precise measurement is critical. Accurate dispensing equipment provided with a graduated glass tube for visual check before discharge is recommended.

Aggregate moisture meters are usually accurate to $\frac{1}{2}$%. They give a reading of free water in the aggregates and can be connected to automatic equipment that makes appropriate adjustments in the quantity of water added to the batch.

MIXING FUNDAMENTALS

Objectives

In any mixing action, the primary objective is to bring water into contact with individual cement particles. The higher the ratio of water to cement, the easier this condition is to bring about. When the water-to-cement ratio is low, a rubbing action is desirable in order to wet the individual particles.

The second step in mixing action is the incorporation of the aggregate into the cement paste in such a manner that all aggregate particles are coated with it, and aggregates are evenly distributed throughout the batch. In mechanical mixing, the blades of the mixer cut, rub, fold, and, in some cases, lift ingredients and allow them to cascade into other portions of the batch.

The principal factors in controlling mixing action in mechanical mixers are capacity, mixing speed, mixing time, and loading sequence of the individual ingredients.

Charging

What effects the order in which materials are added to a mixer has on concrete properties is not definitively known. A few studies on this subject have been published, but the matter remains unsettled.

A given order of introducing concrete mix materials that is successful at one plant may mysteriously cause problems at another. Factors that might cause these variations seem to include type and size of equipment used, variables in the aggregates, cement, water, and admixtures, mix-design proportions, temperatures, and humidity—in other words, almost everything associated with concrete design and production.

Improper order of introduction of mix materials has been cited as one cause for development of cement balls. Cement balls, while not common, do appear occasionally and, if undetected, can seriously undermine the durability and strength of portland cement concrete.

When hot cement is used, cement balls that are not subsequently broken down by the mixing action of the drum are sometimes formed. Truck mixer manufacturers report that they are able to eliminate large cement balls on a number of jobs by altering the charging procedure to eliminate water ponding. When hot cement comes into contact with these pools, flash set takes place and cement balls are formed.

Similar problems arise with the use of hot aggregate or hot water in cold weather batching; these can be alleviated by combining water and aggregates before the introduction of cement.

Mixer blades that are worn, broken, or deformed by accumulations of hardened concrete may be a contributing cause of cement balls; they do not mix the concrete sufficiently enough to break up these formations. Cement balls also seem to occur more frequently in rich, low-slump mixes than in high-slump, high water-content concrete.

Some general rules worth following if cement balls appear are listed here.

1. Begin revolving the drum before charging any materials.
2. Introduce water into the drum first and allow it to clean off blades and drum sides before introducing any other mix ingredients.
3. Continue flow of water into the drum during the entire charging operation and shortly thereafter.
4. Add water at a continuous, slow rate.
5. Avoid the use of unduly hot cement, aggregates, or water.
6. Check the condition of mixing blades.
7. Extend the period of mixing.
8. Refrain from charging the mixer beyond its rated capacity.
9. Mix concrete to low slump first, then add more water to bring it up to the desired level.

The experiences of various producers and researchers vary greatly because of the effect that the order of introduction of mix materials has on concrete strength. The few tests that have been carried out to date are far from definitive.

The 1972 Proposed Revision of ACI Standard 304-73 regarding order of introduction of materials emphasizes "the importance of charging both stationary and truck mixers in a manner to obtain a preblending and ribboning effect as the stream flows into the mixer."

It goes on to say:

"It is preferable that cement be charged with other materials, but it should enter the stream after approximately 10% of the aggregate is in the mixer. When it is necessary to charge cement into truck mixers separately, additional mixing time may be required to obtain desired mix uniformity. Water should enter the mixer first but continue to flow while other ingredients are entering the mixer. Water charging pipes must be of proper design and of sufficient size so that water enters at a point well inside the mixer and charging is complete within the first 25% of the mixing time.

"Admixtures should be charged to the mixer at the same point in the mixing sequence batch after batch. Liquid admixtures should be charged with the water, and powdered admixtures should be ribboned into the mixer with other dry ingredients. When more than one admixture is used, each should be batched separately and they should not be premixed before entering the mixer."

The *ACI Manual of Concrete Inspection* further states:

"Preferably the water should be fed into the mixer over the full period of charging the dry material, beginning just before and ending just after the dry materials are charged. The dry materials should be fed at the same time, preferably so that they will flow in as 'ribbons,' and as rapidly as practical. Loss of materials during charging should not be permitted.

"Hot water (over 140°F) should not be allowed to strike the cement directly, as flash set may occur or cement balls may be formed. Other causes of cement balls are introduction of cement ahead of coarse aggregate, worn mixer blades, hot aggregate or cement, and delayed mixing in truck mixers."

A great deal remains to be learned about the effect of the order of introduction of mix materials on concrete properties. Determination of the effect of charging procedure on slump and more extensive tests of concrete strength and durability are required.

Mixing

All mixers—stationary mixers, truck mixers, or pavers—should be equipped with rating plates giving their designed characteristics. Most truck mixers carry a plate giving capacity and rotation speed of the drum, blades, or paddle. These ratings have been designated as standard by the Truck Mixer Manufacturers Bureau. The size of the mixer unit is indicated by a rating number equal to the maximum capacity in cubic yards when operated as a mixer and also by a higher number when used as an agitator only.

Capacity of truck mixers is usually based on ribbon loading of the ingredients, that is, adding them simultaneously. However, if split loading is used (i.e., each ingredient is added separately), it may be necessary to reduce the size of the batch. The use of rating plates makes possible the realization of maximum payload potential consistent with operating requirements, conditions, and weight laws.

In 1929, Hatt conducted tests on concrete mixers and determined the effect of speed of drum rotation on consistency and strength. Hollister made similar studies on truck mixers in 1932. And in 1954, Walker and Bloem made a further extensive study of truck mixers. These studies showed that the rate of mixing has no consistent effect on the properties of the batch. In actual practice, the higher speed would have the advantage of producing adequate mixing in a shorter time. In all of these studies it was found that time of mixing was the critical factor in controlling uniformity of the concrete. However, in an abridged ACI Summary Paper, Bloem and Gaynor stated:

> "There is little difference between speeds of 4 and 8-10 revolutions pre min. and, in this interval, data confirms earlier researches suggesting that mixing speed has relatively little effect on uniformity. However, the situation changed impressively when speeds in the range from 14 to 18 revolutions per min. were tried. Uniformity improved dramatically!
>
> "This research clearly showed that numerous factors determine whether or not a given batch of concrete will be well mixed—loading or charging procedure, batch size, method of addition of water, drum speed, number of revolutions, and probably several others such as type of materials and mixture proportions."

As early as 1918, D. A. Abrams ascertained that longer mixing increased the strength of concrete. Others conducting research on mechanical mixers have examined the mixing-time factor. Recommended mixing times, in general, have been based on a compromise between optimum physical properties of the concrete and the degree of economy resulting from shorter mixing times.

The average figures in Table 15-A are suitable for evaluating optimum mixing time when there is no provision for testing the uniformity of the

Table 15–A
TIME OF MIXING FOR STATIONARY MIXERS[a]

Capacity	Time of Mixing (min)
1 yd or less	1
2 yd or less	1¼
3 yd or less	1½
4 yd or less	1¾
5 yd or less	2
6 yd or less	2¼

[a] Timing to start after all materials are in the mixer.

batch in the drum, as outlined in ASTM C94, Specification for Ready Mixed Concrete, under "mixing and delivery." This specification provides for reduced mixing time if specified uniformity limits are achieved in less time. The limits include unit weight of air-free mortar, weight of coarse aggregate retained on the No. 4 screen, air content, slump, cement content, and average 7-day strength. They are based on differences in samples taken from the front and rear ends of the load.

The Bureau of Public Roads, in cooperation with 13 state highway departments, has produced a study of the effects of mixing time in 34E dual-drum pavers on the properties of concrete in an effort to determine whether the mixing time usually specified could be shortened. The greater strength was obtained with a mixing time of 60 sec, not including transfer time between drums. Only a very slight reduction in concrete strength resulted from mixing for only 40 sec. Mixing less than 40 sec reduced strength sharply.

Results of studies of the potential cost reduction that can be realized from reduced mixing time are shown in Table 15-B. The base mixing time includes transfer time—about 10 sec.

Construction literature contains more references to the effect of time of mixing on the properties of concrete for truck mixers than for stationary devices. It has long been recognized that methods of loading and the time required to discharge influence the uniformity of the batched product from truck mixers.

The time of mixing, in the case of truck mixers, is indicated by the number of revolutions the drum completes when operated at the speed designated by the manufacturer. ASTM C94 requires that when the concrete is mixed only in a truck mixer loaded to its maximum capacity, 70 to 100 revolutions at the rate of rotation designated by the manufacturer as mixing speed are required to produce the uniformity of concrete re-

Table 15–B
ECONOMY OF SHORTER MIXING TIMES[a]

Base Mixing Time, sec	Production Cost per Cubic Yard	Cost Reduction per Cubic Yard Possible by Decreasing Base Mixing Time to								
		90 sec	75 sec	70 sec	65 sec	60 sec	50 sec	45 sec	40 sec	35 sec
120	$6.69	$1.85	$2.36	$2.51	$2.64	$2.77	$2.98	$3.08	$3.17	$3.25
90	4.84		0.51	0.66	0.79	0.92	1.13	1.23	1.32	1.40
75	4.33			0.15	0.28	0.41	0.62	0.72	0.81	0.89
70	4.18				0.13	0.26	0.47	0.57	0.66	0.74
65	4.05					0.13	0.34	0.44	0.53	0.61
60	3.92						0.21	0.31	0.40	0.48
50	3.71							0.10	0.19	0.27
45	3.61								0.09	0.17
40	3.52									0.08

[a] From *Public Roads*, April 1960.

quired by the specification. Because of the need to keep track of the number of revolutions, all mixers should be equipped with a revolution counter.

Overmixing

Grinding. Mixing of concrete is a grinding process. Prolonged mixing can produce excessive grinding and reduce the size of aggregates.

Table 15-C shows variations in aggregate size after 5 min of mixing. The data indicates that mixing produces a decrease in the quantity of coarse aggregate, 1½ to ⅜ in., causing a corresponding increase in each of the quantities of finer sizes except the No. 30 to No. 50 mesh sizes.

The amount of grinding that will occur during the mixing of concrete depends mainly on the hardness of the aggregate and the degree of mixing. The data in Table 15-C is based on the use of a reasonably hard dolomite coarse and fine aggregate mixed for 5 min in an open tub mixer. More aggregate breakdown would occur with softer aggregate and longer or more violent mixing.

The increase in fine aggregate that occurs during prolonged mixing can have a significant effect on properties of concrete. Such increase will cause a decrease in slump or an increase in the mixing water required to maintain the original slump.

The grinding action during mixing of concrete also generates heat. Thus, prolonged mixing can raise the temperature of the concrete thereby further increasing the amount of water required to maintain consistency.

Table 15–C
VARIATIONS IN AGGREGATE SIZE AFTER FIVE MINUTES OF MIXING

	Percentage of Size Group		Percentage of Total Aggregate	
Size Group	Decrease After Mixing	Increase After Mixing	Decrease After Mixing	Increase After Mixing
1½" to ¾"	3.5		0.7	
¾" to ⅜"	15.0		3.1	
⅜" to No. 4		6.0		1.2
No. 4 to No. 8		13.0		0.9
No. 8 to No. 16		7.3		0.5
No. 16 to No. 30		7.3		0.5
No. 30 to No. 50	14.7		1.0	
No. 50 to No. 100		4.5		0.3
Minus No. 100		41.0		1.4

Slump. Prolonged mixing decreases the slump. This decrease is caused by the combined effect of the grinding action of mixing, the hydration of the cement, evaporation, and aggregate absorption.

Table 15-D shows the effect of prolonged mixing on slump. It is based on tests made using a relatively hard sand and gravel aggregate with a concrete temperature of about 70°F. The results show only a slight reduction in the slump during the first 10 min of mixing, but show a rapid reduction beyond this time. The reduction in slump would be more pronounced if a softer aggregate, one that would grind more rapidly, were used, or if the concrete were at a higher temperature, with the subsequent accelerated cement hydration.

Air Content. Prolonged mixing causes a reduction in the air content and slump of air-entrained concrete (as shown in Table 15-E). The reduction in air content may result from an increase in very fine particles or simply from an increase in the ratio of air escape to foam generation in the later

Table 15–D
EFFECT OF PROLONGED MIXING ON THE SLUMP OF CONCRETE

Mixing Time, min	Slump, in.
2	5.0
5	4.9
10	4.7
15	4.1
30	3.4
60	2.5

Table 15–E
AIR-ENTRAINED, HIGH SLUMP CONCRETE

Mixing Time, min	Concrete Sample	Slump, in.	Percent Air (Roll-a-Meter)	Compressive Strength	
				PSI	Percent of Initial
6	Initial	5.6	5.8	4445	100
50	NR	1.6	4.6	4650	104.8
50	R	6.1	3.8	3523	79.4
100	NR	0.4	2.0	4954	111.8
100	R	5.4	2.3	3131	70.5
150	NR	0.0	Too dry to test		
150	R	5.6	2.0	2470	55.6

stages of mixing. Entrained air improves the workability of concrete. So reduction in air content contributes to reduction in slump.

Table 15-E shows the effects of prolonged mixing on strength as well as on air content and slump. The initial sample tested contained 489 lb of cement per cu yd and 275 lb of water per cu yd. "NR" in the table means not retempered; "R" means retempered. Note that in the retempered batches, the slump has been restored, but strength was drastically reduced and air content did not increase.

Retempering

Application. Table 15-E gives fair warning of the possible dangers of overmixing and retempering. Both are generally considered bad practice. Yet, every concrete worker has been faced at some time or other with a borderline case, when the advantages of retempering might far outweigh any possible disadvantages. Currently, the hardships and greatly increased costs that can result from an overstrict observance of the no-retempering rule are receiving some well-deserved study. Considerable research is underway by various agencies, the tentative results of which are already showing that the effects of retempering may not be so serious and that a more lenient interpretation of the total prohibition rule may be possible.

Drastic slump reductions can be attributed broadly to prolonged mixing, excessive standby time, or elevated temperatures. The usual precaution is to specify that discharge take place within 90 min after mixing. Forty-seven state highway departments impose this rule. Several states require discharge within 30 min when Type III cement is used or when the temperature is 90°F or above; they permit the 90-min maximum with Type I cement at lower temperatures. These practices generally fall in line with the ASTM standard specification (C94-68) for ready-mixed concrete that requires concrete delivered in a truck mixer or agitator to be discharged within 90 min or 300 revolutions, whichever comes first, after the addition of mixing water to the cement and aggregates or after the addition of cement to the aggregates.

While this specification apparently gives reasonably wide operating freedom, the occasions when it becomes restrictive are unfortunately all too many. Long hauling distances, breakdowns, and unusual placing requirements can give rise to conditions in which retempering is highly advantageous. Such factors as heavy reinforcement and the type and location of the structure often require that concrete be placed at a slow rate to avoid segregation. If structures at a project require small quantities of concrete and excessively long hauls are involved, the use of ready mix may become prohibitively costly. If specifications are followed, concrete is either wasted or delivered in small batches at increased contractor and client cost. If they are not followed, additional water must be added to the concrete to restore slump and provide sufficient workability for proper placing and compaction. A conventional objection is that additional water means higher water-cement ratio and lower strength and durability.

On many better-controlled jobs, it is not uncommon to permit concrete in truck mixers to be brought to desired slump upon arrival at the job site, but without the use of any additional water. The approved method of accomplishing this is by deliberately batching on the dry side and then adding the required amount of mixing water after delivery. This procedure is outlined in more detail in ASTM C94, sec. 9.7. This practice is not retempering. The Bureau of Reclamation points out that "for ideal conditions of control, the initial charge of water would be such as to bring the mixture to a plastic state at slightly less than the desired slump. The mixture would then be brought to the desired consistency by the addition of tempering water."

Use of Admixtures. Available retarding admixtures are capable of delaying the hardening of concrete so that it can be vibrated or finished much later than normally without adversely affecting strength development. It is logical to expect that these same admixtures might provide additional benefits in ready-mixed concrete production by reducing the loss of slump due to prolonged mixing, long hauls, or warm weather. However, it is possible to experience more slump loss with extended mixing times than might be expected without the addition of an admixture.

ACI Recommendation. The 1972 Proposed Revision of ACI 614-59 makes this recommendation:

"Provided the design water-cement ratio is not exceeded, small increments of retempering water may be added to mixed batches to obtain the desired slump. However, the production of concrete of excessive slump or addition of water in excess of the design water-cement ratio to compensate for slump loss resulting from delays in delivery or placing should be prohibited. . . . the water required for proper concrete consistency (slump) is affected by such things as amount and rate of mixing, length of haul, time of unloading, and ambient temperature conditions. In cool weather or for short hauls and prompt delivery, such problems as loss or variation in slump, excessive mixing water requirements, and discharge, handling and placing problems rarely exist. However, the reverse is true when rate of delivery is slow or irregular, haul distances are long, and weather is warm. Additions of water to compensate for slump loss should not exceed that necessary to compensate for a 1-in. (2.5 cm) loss in slump, nor should the design maximum water-cement ratio be exceeded. Loss in workability during warm weather can be minimized by expediting delivery and placement, by controlling the mix temperature, and when appropriate by the use of retarders. When feasible all mixing water should be batched at the central plant. However, in hot weather it is frequently desirable to withhold some of the mixing water until the mixer arrives at job. Then with the remaining required water added, an additional 30 revolutions at mixing speed is required to adequately incorporate the additional water into the mix. When loss of slump or workability cannot be offset by these measures, complete mixing should be performed on the job, using centrally dry batched materials."

In summary, retempering should be avoided through careful truck scheduling, proper mixer maintenance, and careful job planning. If delays are unavoidable, use of admixtures or retempering according to the ACI recommendations may be considered.

Discharge

All types of mixers should be capable of ready discharge of concrete of the lowest slump able to be consolidated by vibration. In many massive or unformed placements, modern vibrators will readily consolidate concrete of a 1-in. slump. Separation of coarse aggregate from the mortar, which commonly results when concrete is discharged from most plant and transit mixers, should be avoided by arrangement of the discharge so concrete will fall vertically, not diagonally, into whatever container is to receive it.

Blade arrangement and discharge mechanism of all types of mixers—including agitating, shrink, and transit mixers—should be such that throughout the discharging operations the aggregate is well distributed from coarse to fine. Should the last fraction of the batch contain an excessive amount of coarse aggregate, this portion should be retained and mixed with the succeeding batch. Batch size in this case should be reduced by an amount corresponding to the quantity withheld. In some truck mixers this type of separation may be alleviated by reversing the direction of rotation for the last 10 or 12 revolutions prior to discharge. Coarse aggregate is considered in objectionable excess when any larger than ¾ in. exceeds the amount of that size batched by 15%.

Mixing Difficulties

If split loading of truck mixers is used, and the sand is inadvertently put in first, it will stick to the shell of the drum and will not mix properly. No amount of mixing will correct it. When such a mix is discharged, balls of unmixed sand appear throughout the batch. Sometimes a condition similar to that producing cement balls occurs when a little of the portland cement is carelessly added ahead of the other ingredients. These conditions, originating at loading time, must be avoided.

Evaluating Mixers

The efficiency of all types of mixers can be evaluated by means of within-batch uniformity measurements. Such measurements as slump, air content, unit weight, strength, and cement content by centrifuge separation have all been used for studying the uniformity of concrete from stationary mixers, pavers, and truck mixers.

A program of tests to establish within-batch uniformity is valuable for checking on required amount of mixing, establishing capacities of mixers, detecting blade wear, and evaluating batching sequences. Some of the tests that have been used are easily conducted with ordinary field-testing equipment; others involve refined laboratory techniques and the meticulous care of skilled technicians.

Evaluation tests that can be made in the field are slump, unit weight, air determination, washout tests for checking the grading of aggregate, and casting cylinders for laboratory compression tests. For all of these, the concrete should be sampled during discharge from the mixer by passing a shovel or bucket through the entire stream of concrete. Three or more 1-cu ft samples from every batch should be taken. The sampling procedure should comply with the ASTM Standard Method of Sampling Fresh Concrete (C172).

Freshly sampled concrete should be remixed with a shovel just enough to correct any segregation caused by sampling. A slump test should be made immediately after remixing. The remainder of the sample should be used to make a unit-weight test, a wash test, and three concrete cylinders for strength tests.

The slump test, the unit-weight test, and the cylinders for strength testing should all be made in accordance with ASTM standard methods.

The wash test consists of washing a sample of concrete over a reinforced No. 100 sieve topped by a No. 4 sieve. The concrete sample is weighed before washing, and the stone and sand portions should be weighed separately after washing. The materials are dried overnight, and sieve analyses are made of both stone and sand samples. The entire coarse aggregate sample is used in the sieve analysis. A 500-g sample previously dried over a hot plate is used for sieve analysis of the sand. Concrete uniformity standards, as designated by ASTM, are listed in Table 15-F.

Maintenance of Mixers and Blade Wear

A mixer, particularly its blades, must be clean and in good mechanical condition in order to ensure satisfactory mixing action—the complete coverage of aggregate surface by the cement-water paste. Any accumu-

Table 15–F
REQUIREMENTS FOR UNIFORMITY OF CONCRETE (ASTM C94)

Test	Requirement[a]
Weight per cu ft, calculated to an air-free basis	1.0 lb
Air content, based on volume of concrete	1.0%
Slump	
If average slump is 4 in. or less	1.0 in.
If average slump is 4 to 6 in.	1.5 in.
Coarse aggregate content, portion by weight of each sample	
retained on No. 4 sieve	6.0%
Unit weight of air-free mortar, based on average for all comparative	
samples tested	1.6%
Average compressive strength at 7 days for each sample,[b] based on	
average strength of all comparative samples tested	7.5%

[a] Expressed as maximum permissible difference in results of tests of samples taken from two locations in the concrete batch.
[b] Not less than 3 cylinders will be molded and tested from each of the samples.

lation of material on the inner surface of the drum will affect this mixing action adversely. Leaky water valves are also a source of detrimental effect on concrete uniformity, particularly during the periods between batches.

The object of all mixing is uniformity throughout the entire course of a batch. In addition to mixer maintenance, this requires following manufacturers' recommendations for capacities, drum/blade speeds, and length of mixing time. For truck mixers, proper loading procedures should be employed to prevent segregation.

Inefficient mixing can often be traced directly to blade problems. If blades are worn or encrusted with hardened cement, mixer operations will be adversely affected. Worn blades should be replaced immediately, and accumulations of hardened concrete should be removed when first noticed.

MIXER INSTALLATIONS

Mechanical mixers are available in many types and sizes for a number of specific applications. They should never be loaded above their rated capacities* or operated at speeds other than those for which they were designed. If increased output is desired, larger capacity mixers or additional lower capacity machines should be employed.

Mixer plants are basically of two varieties: (1) portable, including conventional portable mixers and paving mixers, and (2) permanent, plants established at paving sites, mass-concrete installations, and ready-mix plants. Truck mixers, also considered mixing plants in a sense, are covered in detail in *Basic Concrete Construction Practices*, Chapter 5.

Mixers for Portable Plants
Small Portable Mixers. If only a small amount of concrete is needed, small tilting or nontilting mixers can be used. These machines are available in various sizes for mixing batches ranging from 2 to 16 cu ft.

Tilting mixers are generally charged by hand shoveling; they have no batchmeter or other timing device. It is not uncommon for a mix to be discharged as soon as the workman has finished shoveling. Although these small tilting machines are low in cost and generally do an adequate job for small concrete batches, they should not be used when controlled concrete is specified.

---

* Rated capacity is usually indicated on a plate affixed to the equipment.

Nontilting mixers are well suited for producing quality concrete. They can be purchased equipped with water-measuring tanks and with batchmeters. These tanks measure water by volume and automatically charge the mixer with this preselected amount. The batchmeter controls mixing time by locking the discharge mechanism until a set mixing time has elapsed. The operator is alerted by a bell when a mix is ready to be discharged. The meter automatically restarts when the mixer is again charged.

Paving Mixers. Paving mixers, where used, are generally either the single-drum 27E mixer or the two-drum 34E mixer (the numbers refer to capacity in cubic feet); a larger three-drum mixer has also been developed.

Although paving mixers were used extensively in the past, they have been supplanted considerably by central-mixing plants. It is now much more common for concrete to be mixed at a central plant, delivered to a job site, and dumped directly onto a subgrade, subbase, or special belt or hopper type spreader ahead of the paver. This change is due primarily to greater concrete quality control and higher production rate these plants provide. Many of them are highly sophisticated and automated.

Permanent Installations

Stationary Mixers. Stationary mixers are used for central mixing in ready-mix plants or directly on job sites. Many are equipped with timing devices that automatically lock the discharge mechanism until a preset mixing time has elapsed. Water is first introduced into the mixer just prior to the addition of the solid materials; it continues to be added until shortly after all materials are in the mixer. Mixing actually begins when all solid materials are in the drum or pan and ends with discharge.

There are four basic types of stationary mixers: tilting and non-tilting-drum types, vertical-shaft pan type, and horizontal-shaft type.

Tilting mixers consist of a rotating drum mounted on trunnions that allow the drum to be tilted for charging and discharging. The nontilting mixer discharges through a chute that can swing inside the drum. Pickup blades within the drum deposit the concrete onto the chute, which carries it to another outside guide chute or hopper from which it is directed to the haulage unit.

Tilting mixers have some advantages over the nontilting kind.

1. They can handle much larger aggregate. (Nontilting mixers are normally limited to mixing concrete with aggregates smaller than 3 in. in size.)

2. They discharge much more rapidly because the concrete is simply dumped from the mixer.

On the other hand, nontilting mixers are less expensive and have lower installation and operating costs.

The vertical-shaft pan-type mixer is a more recent development and is becoming more and more popular in the concrete industry. It consists of shaft-mounted paddles or blades rotating in an open-top pan; the pan may also rotate. If it does, it is in the direction opposite that of the paddles. Most rotating pan-type mixers have some of the mixing arms combined into three- or four-armed stars that rotate about their centers as well as turning about the shaft. It is essential that the blades meet the concrete with the greatest possible speed and, therefore, force, to ensure a good mix. It is also important that the mixing pan be completely utilized.

The rotating pan-type mixer requires less time to produce a finished mix, and it mixes more intensely than conventional drum-type mixers. Output is increased, but higher wear and more frequent maintenance result. The principal advantage of this mixer is that it mixes low slump concretes efficiently and higher slump concretes adequately. These pan mixers can turn out concrete of near-zero slump for masonry units or pipe, low-slump concrete for prestressed or precast products, and higher-slump concrete for ready-mix uses.

A vertical-shaft mixer requires less height and can usually be installed simply and inexpensively. Because of their smaller size, two or more units can often be installed in the same area one conventional unit requires.

A possible disadvantage of these mixers is their somewhat greater tendency to wear.

The horizontal-shaft mixer consists of a power-driven spiral blade or paddle operating inside a horizontal drum. Tests have shown it to be highly efficient.

Ready-Mix Plants. There are two types of ready mixed concrete plants: truck mixing and central mixing.

The *truck-mixing plant* is simpler because it does not have a stationary mixer in the plant. Materials at the plant are handled, stored, and batched. These ingredients for concrete are mixed completely in a truck mixer, either at the plant, in transit, or on arrival at the job site.

Truck-mixing plants generally are less expensive and have lower operating costs than do central-mixing plants. They are more numerous and better located. Truck-mixing plants are capable of a higher production rate from equal-size batchers. Also, for long hauls, the truck mixer can be charged dry; water is added and mixing done directly on the job site.

The *central-mixing plant* contains a stationary plant mixer. Batchers discharge into this mixer, and the resultant concrete is discharged into a haulage unit—an agitator (a truck mixer with a revolving drum that agitates the materials) or a special agitating or nonagitating dump truck.

Some central-mixing plants are equipped to bypass the central mixer and charge a truck mixer directly from the weigh batcher, thus combining the functions of both types of plants. Such combination plants usually require relatively long chutes since the batcher is located above the plant mixer. These long dry-batch chutes can be a source of trouble. Severe dusting of the cement may occur due to the long drop, and a paste of cement and moist sand may build up rapidly on the sides and cover of the chute. These problems are controlled in many modern low-profile plants where batchers are located at ground level. Belt conveyors lift materials directly from the weigh batchers on to a short mixer-charging chute.

In some central-mixing plants, mixers are used to mix the materials partially, the mixing being completed in a truck mixer. This practice is referred to as *shrink mixing* (the incomplete mixture is called preshrunk concrete) because the volume of materials is less than it would be if the materials were dry charged into the truck mixer. For example, the materials going into a 6-yd batch of concrete will, before mixing, have a combined volume considerably greater than 6 cu yd. Shrink mixing reduces this volume appreciably, although not as much as complete mixing will. It thereby allows more concrete to be charged into a truck mixer than does dry charging. Also, build up of materials on the forward end of the mixer drum is reduced.

The main advantage of central-mixing plants is their better control of concrete. The chief disadvantage is that the operation is less flexible if special materials are required or if concrete must be transported long distances in hot weather. Central-mixing plants are used in about 25% of ready-mixed concrete plants and 70% of all highway paving jobs. Many have completely automatic batching devices equipped with interlocking mechanisms to ensure uniform concrete. Even in central-mixing plants lacking automation, the centralization of responsibility for quality concrete is an important advantage. Usually only one man—the plant operator—is responsible for maintaining control of concrete consistency.

When the concrete is completely mixed in the plant mixer and hauled in a revolving drum truck, payloads average 15 to 20% larger. If hauling distances are not too long, central-mixing plants use regular dump trucks; these vehicles are much less expensive than truck mixers. Dump trucks are being used increasingly in paving operations. Modern spreaders are designed to receive concrete from a rear or side dump truck, and these trucks are also commonly used to dump concrete onto subgrades or subbases directly ahead of a paver. Dump trucks are best suited for hauling relatively low-slump concrete; if slump is too high, segregation may occur.

Several manufacturers are now offering mobile and portable ready-mix plants that can either be truck-mixing or central-mixing types. Mobile plants are constructed such that their main components can be disassembled quickly and hauled by trucks. Reassembly at another location can usually be accomplished rapidly and without much trouble. Portable plants can be moved easily, but with somewhat more difficulty than the mobile plant.

On-Site Central-Mixing Paving Plants. Many contractors are using central-mixing plants installed at the paving sites. These plants include automatic or semiautomatic batchers and large stationary mixers with 6- to 12-cu yd capacities. The different materials, including water, are batched into mixers simultaneously (ribbon loaded), and the largest mixers take only 15 to 30 sec to complete batching. The discharge mechanism is time locked to ensure an adequate mixing period. High-volume production can be attained with this kind of operation.

Concrete is transported to the paving machine in special trucks equipped with or without agitators. The truck body has an open top and is shaped for tilt discharge through a control gate located at the rear or side. Vibrators are sometimes used to facilitate unloading. Ordinary dump trucks are especially common when spreader equipment is used in placing.

Mass Concrete Plants. Mass concrete plants are set up on the site to produce concrete for particular jobs, always large and important projects such as dam construction. Project specifications determine the nature of the mixing plant and the objectives to be achieved in batching. Most likely a new plant will be installed, although it is sometimes possible and more economical to set up a previously used plant that has been moved in and modified to suit the requirements of the job.

Usually a tilting mixer is found in every modern mass concrete plant. The nontilting mixer is unsuited for use in these plants because it cannot discharge rapidly enough or mix concrete containing large aggregates. Sometimes two or more mixers that discharge into a common collecting hopper are installed. Hoppers may be compartmented to receive several different mixes simultaneously.

GLOSSARY

absorption. The process by which a liquid is drawn into and tends to fill permeable pores in a porous solid body.

accelerator. A substance that, when added to concrete, mortar, or grout, increases the rate of hydration of a hydraulic cement, shortens the time of set, or increases the rate of hardening or strength development.

admixture. A material other than water, aggregates, and hydraulic cement, used as an ingredient of concrete or mortar, and added to the batch immediately before or during its mixture.

aggregates. Granular material such as natural or manufactured sand, gravel, crushed gravel, crushed stone, and air-cooled iron blast-furnace slag which, when bound together into a conglomerated mass by a matrix, form concrete or mortar.

air content. The volume of air voids in cement paste, mortar, or concrete, exclusive of pore space in aggregate particles, usually expressed as a percentage of total volume of the paste, mortar, or concrete.

air-entrained concrete. Concrete containing an air-entraining agent that causes millions of minute bubbles of air to be trapped within the concrete. Air-entrained concrete is more resistant to freeze-thaw cycles than non-air-entrained concrete.

air-entraining cement. Cement containing an air-entraining agent in such amount as to cause the resulting concrete to contain entrained air within specified limits. Designated by the letter "A" after "Type," such as Type IA and Type IIA, when describing cements.

air void. A space in cement paste, mortar, or concrete filled with air; an entrapped air void is characteristically 1 mm or more in size and irregular in shape; an entrained air void is typically between 10 and 1000 microns in diameter and spherical or nearly so.

alkali. A water-soluble salt of one of the alkali metals, principally sodium and potassium; usually expressed in chemical analysis of portland cements as the oxides Na_2O and K_2O.

ambient. Completely surrounding, (e.g., ambient air).

ASTM. American Society for Testing and Materials.

autoclave. A pressure vessel in which an environment of steam at high pressure may be produced. Used in the curing of concrete products and in the testing of hydraulic cement for soundness.

axial. Situated in an axis or on the axis.

ball mill. A horizontal rotating cylinder containing large steel balls capable of grinding materials.

batching. Measuring the ingredients for a batch of concrete or mortar by weight or volume and introducing them into the mixer.

batch meter. A device that controls mixing time by locking the discharge mechanism so that the mixer cannot be discharged until the preset mixing time has expired.

beneficiation. Improvement of the chemical or physical properties of a raw material (e.g., aggregates) or intermediate product by the removal of undesirable components or impurities.

bleeding. The autogenous flow of mixing water from newly placed concrete or mortar, caused by the settlement of the solid materials within the mass; also called water gain.

bulking. Increase in the bulk volume of a quantity of sand in a moist condition over the volume of the same quantity dry or completely inundated.

calcareous. Containing calcium carbonate or, less generally, containing the element calcium.

capillary flow. Flow of moisture through a capillary pore system, such as in concrete.

capillary space. In cement paste, any space not occupied by anhydrous cement or cement gel. (Air bubbles, whether entrained or entrapped, are not considered to be part of the cement paste).

cast-in-place. Mortar or concrete that is deposited in the place where it is required to harden as part of the structure, as opposed to precast concrete.

cement, portland. The product obtained by pulverizing clinker consisting essentially of hydraulic calcium silicates; usually containing calcium sulfates as an interground addition.

central mix. Concrete that is completely mixed in a stationary mixer before being transported to the site in agitating or nonagitating dump trucks.

charging. Feeding materials into the mixer.

clinker. A partially fused product of a kiln, which is ground to make cement; also other vitrified or burnt material.

coefficient of thermal expansion. Change in linear dimension per unit length or change in volume per unit volume per degree of temperature change.

compressive strength. The measured maximum resistance of a concrete or mortar specimen to axial loading expressed as force per unit of cross sectional area.

concrete. An intimate mixture of cement, aggregates and water (and perhaps small amounts of other materials), which will or has hardened into a rocklike mass.

consistency. The relative mobility or ability of freshly mixed concrete or mortar to flow.

contraction joints. A plane, usually vertical, separating concrete in a structure or pavement, at a designated location such as to interfere least with performance of the structure, yet such as to prevent formation of objectionable shrinkage cracks elsewhere in the concrete.

cumulative batchers (multiple batchers). A weighing system in which several materials are weighed one after another in the same hopper using a single-scale lever system.

curing. Maintenance of humidity and temperature of concrete during some definite period following placing, casting, or finishing to assure satisfactory hydration of the cementitious materials.

dehydrate. Take moisture from.

deleterious. Injurious or harmful.

deterioration. A worsening or lowering in quality.

diatomaceous earth. A friable earthy material composed of nearly pure hydrous amorphous silica (opal) and consisting essentially of the frustules of the microscopic plants called diatoms.

dilation. Enlargement; expansion.

dispersing agent. An addition or admixture capable of increasing the fluidity of pastes, mortars, or concrete by reduction of interparticle attraction.

diametral. 1. Of a diameter. 2. Forming a diameter.

drying shrinkage. Contraction caused by moisture loss.

durability. The ability of concrete to resist weathering action, chemical attack, abrasion, and other conditions of service.

efflorescence. A deposit of salts, usually white, formed on a surface, the substance having emerged from below the surface.

entrained air. Microscopic air bubbles intentionally incorporated in mortar or concrete during mixing.

entrapped air. Air voids in concrete that are not purposely incorporated in the batch.

false set. The rapid development of rigidity in a freshly mixed portland cement paste, mortar, or concrete (generally caused by dehydration of gypsum) without the evolution of much heat, which rigidity can be dispelled and plasticity regained by further mixing without the addition of water; "premature stiffening," "hesitation set," "early stiffening," and "rubber set" are terms referring to the same phenomenon, but "false set" is the preferred designation.

fines. As concrete is troweled, a mixture of cement paste and the fine aggregate is brought to the surface to help seal any small holes. This mixture is called the fines.

flash set. The rapid development of rigidity in a freshly mixed portland cement paste, mortar, or concrete, usually with the evolution of considerable heat, which rigidity cannot be dispelled nor can the plasticity be regained by further mixing without addition of water; also referred to as "quick set" or "grab set." It is generally caused by the combination of excessively hot ingredients.

flexural strength. A property of a solid that indicates its ability to withstand bending.

freeze-thaw cycle. The cycle of freezing, melting, freezing, melting. In most northerly climates, pavement (and other concrete construction) will freeze in the night and thaw during the day when the sum hits the surface.

friable. Able to be broken into small particles by finger pressure.

frustum. The part of a conical solid left after cutting off a top portion by a plane parallel to the base.

gel. Matter in a colloidal state that does not dissolve but remains suspended in a solvent from which it fails to precipitate without the intervention of heat or of an electrolyte.

Gillmore needle. A device used to determine time of setting of hydraulic cement.

gradation (or particle size distribution). The distribution of particles of granular material among various sizes.

grout. Mixture of cementitious material and fine aggregate to which sufficient water is added to produce a pouring consistency without segregation of the constituents.

grouting. The process of filling with grout.

gypsum. A mineral having the composition calcium sulfate dihydrate ($CaSD_4 - 2H_2O$).

hardener. A chemical (including certain fluosilicates or sodium silicate) applied to concrete floors to reduce wear and dusting.

heat of hydration. Heat evolved by chemical reactions with water, such as that evolved during the setting and hardening (hydration) of portland cement.

homogeneous. Of the same kind, similar.

honeycomb. Voids left in concrete due to failure of the mortar to effectively fill the spaces among coarse aggregate particles.

humus. Soil made from decayed leaves and other vegetable matter, containing plant food.

hydration. Formation of a compound by the combining of water with some other substance; in concrete, the chemical reaction between cement and water.

hydraulic cement. A cement that is capable of setting and hardening under water due to interaction of water and the constituents of the cement.

hydraulic pressure. Pressure transmitted by or due to the weight of a vertical column of water.

igneous. Formed by volcanic action or great heat (e.g., igneous rock).

inhibitor. A material used to reduce corrosion of metals imbedded in concrete.

inundation batching. Measurement of the volume of sand immersed in water as determined by the displacement of the water. This method eliminates errors in measurement of volume caused by bulking of sand due to moisture.

inverse. Reversed in position, direction or tendency: inverted.

kiln. A furnace or oven for drying, charring, hardening, baking, calcining, sintering, or burning various materials.

laitance. A layer of weak and nondurable material containing cement and fines from aggregates brought by bleeding water to the top of overwet concrete.

lignite. A dark-brown type of soft coal in which the texture of wood can be seen.

line of traverse. A line that passes across, through or over.

loam. A rich, dark soil composed of sand, clay, and some organic matter.

metamorphic. In rock, a change in structure brought about by pressure, heat, chemical action, and so on.

micron. A unit of length: one-thousandth of a millimeter or one-millionth of a meter.

mineral admixture. Admixtures containing inorganic substances such as pozzolans that are used to reduce cement requirements, heat buildup and expansion of concrete.

mix water. The water in freshly mixed sand-cement grout, mortar, or concrete, exclusive of any absorbed by the aggregate (i.e., water considered in the computation of the net water-cement ratio).

modulus of elasticity. The ratio of normal stress to corresponding strain for tensile or compressive stresses below the proportional limit of the material; referred to as "elastic modulus," "Young's modulus," and "Young's modulus of elasticity"; denoted by the symbol E.

modulus of rupture. A measure of the ultimate load-carrying capacity of a beam and sometimes referred to as "rupture modulus" or "rupture strength." It is calculated for apparent tensile stress in the extreme fiber of a transverse test specimen under the load which produces rupture.

monolithic construction. A process of casting or erecting a body of plain or reinforced concrete as a single integral mass with no joints other than construction joints.

mortar. A mixture of cement paste and sand.

mortar cube. A 2-in. cube composed of 1 part cement and 2.75 parts sand by weight. The sand is a graded Ottawa silica sand, and the water requirement is determined experimentally, (ASTM C109, Section 9).

mud-pumping. If the subbase of the roadbed is not properly drained, the weight of heavy vehicles passing over the slab of pavement above causes a see-saw motion that sets up pressures that pump the subbase from under the slab. This is called mud-pumping.

NCMA: National Concrete Masonry Association.

neat cement paste. Cement mixed with water, no aggregate.

NRMCA. National Ready Mixed Concrete Association.

optimum. The best or most favorable degree, condition.

organic materials. Substances that came orginally from living plants or animals. These include fats, petroleum products, sugars, coal, and wood. All of these contain the chemical element carbon.

pan mixer. Mixer consisting of a horizontal pan or drum in which mixing is accomplished by means of the rotating pan or fixed or rotating paddles or both. Rotation is about a vertical axis.

paver. Mobile equipment used to place and finish concrete pavement. The equipment, which runs on wheels or crawler treads, usually consists of a mixer, screws or conveyors for distributing the concrete, screeds, and power trowels. Several machines that travel over the roadbed in succession are often used. This assemblage of equipment is called a paving train.

permeability. A characteristic or substance that allows the passage of fluids.

petrographer. An expert in dealing with the description or classification of rocks.

poise weights. The movable weights on the scale's weightbeam.

Poisson's ratio. The ratio of transverse (lateral) strain to the corresponding axial (longitudinal) strain resulting from uniformly distributed axial stress below the proportional limit of the material; the value will average about 0.2 for concrete and 0.25 for most metals.

popout. The breaking away of small portions of a concrete surface due to internal pressure that leaves a shallow, typically conical depression.

porosity. The ratio, usually expressed as a percentage, of the volume of voids in a material to the volume of the material, including the voids.

portland cement. See cement, portland.

pozzolan. A siliceous or siliceous and aluminous material that in itself possesses little or no cementitious value but will, in finely divided form and in the presence of moisture, chemically react with calcium hydroxide at ordinary temperatures to form compounds possessing cementitious properties.

ppm. Parts per million number of units of measurement (inch or ounces, gallons, pounds, or grams) of the substance present in a million of the same measuring units, for instance 5 lb of salt per million pounds of water = 5 ppm.

prestressed concrete. Concrete in which internal stresses of such magnitude and distribution are introduced that the tensile stresses resulting from the service load are counteracted to a desired degree; in reinforced concrete the prestress is commonly introduced by tensioning the tendons.

psi. Abbreviation for pounds per square inch.

reinforced concrete. Concrete containing reinforcement and designed on the assumption that the two materials act together in resisting forces.

resin. Any of a class of nonvolatile, solid or semisolid organic substances (copal, mastic, for example) obtained directly from certain plants as exudations or prepared by polymerization of simple molecules, and used in medicine and in the making af varnishes and plastics. It often has a high molecular weight and a tendency to flow under stress. It usually has a softening or melting range and usually fractures conchoidally.

retarder. An admixture that delays the setting of cement paste, and hence of admixtures such as mortar or concrete.

retempering. The addition of water and remixing of concrete or mortar that has started to stiffen.

retrogression. Moving backward, becoming worse.

revetment. A concrete facing to protect a wall—usually from erosion by water.

ribbon charging. Putting materials in the mixer simultaneously so that they will flow in as ribbons and as rapidly as practical.

saturated. To be thoroughly soaked to capacity.

scaling. Local flaking or peeling away of the near-surface portion of concrete or mortar.

sedimentary. Formed by the deposit of sediment, as rocks.

segregation. Differential concentration of the components of mixed concrete resulting in nonuniform distribution.

sequence charging. Putting materials in the mixer one at a time.

set. The condition reached by a cement paste, mortar, or concrete when it has lost plasticity to an arbitrary degree, usually measured in terms of resistance to penetration or deformation: initial set refers to first stiffening; final set refers to attainment of significant rigidity.

setting time. The time required for a neat cement paste or mortar to attain a certain degree of rigidity as measured by the Gillmore or Vicat needles.

shrink-mixed concrete. Ready-mixed concrete mixed partially in a stationary mixer to reduce the volume that the concrete will occupy. The batch is then charged into a truck mixer for completion of the mixing en route to the job site.

skip. A scooplike hopper usually used with small portable mixers or paving equipment for charging aggregates and cement into the mixed drum. The skip lowers to accept materials and is raised to allow the materials to slide into the drum.

slabjacking. Grout is forced under the slab, returning it to its original place and stabilizing the subbase.

slip-form paver. A machine that moves along a prepared roadbed forming a ribbon of pavement as it goes. The consistency of the concrete is such that it does not change form after being placed. The slipform paver finishes the surface of the concrete as it goes along.

soundness. The freedom of a solid from cracks, flaws, fissures, or variations from an accepted standard: in the case of a cement, freedom from excessive volume change after setting.

spacing factor. An index related to the maximum distance between air voids.

slurry. A mixture of water and any finely divided insoluble material, such as portland cement, slag, and soil in suspension.

specific gravity. The ratio of the mass of a unit volume of a material at a stated temperature to the mass of the same volume of gas-free distilled water (for liquids and solids) or air or hydrogen (for gasses) at a stated temperature.

specific surface. The surface area of particles contained in a unit weight or absolute unit volume of a material.

split loading. Method of charging a mixer in which the solid ingredients do not all enter the mixer together. Cement and sometimes different sizes of aggregate may be added separately.

stearate. A salt of stearic acid, a white fatty acid such as tallow.

subbase. A layer in a pavement system between the soil prepared to support the pavement and the pavement itself.

sulfate attack. Harmful or deleterious chemical or physical reaction or both between sulfates in soil or ground water and concrete or mortar, primarily the cement paste matrix.

tamped concrete pipe. Pipe formed by compacting freshly placed concrete with repeated blows.

tarebeam. Beam used to neutralize the weight of the hopper, container for the water or the aggregates.

tensile strength. Maximum stress that a material is capable of resisting under axial tensile loading, based on the cross-sectional area of the specimen before loading.

thermal stress: Internal stress caused by heat.

tilting mixer. A horizontal axis mixer, the drum of which can be tilted. The materials are fed in when the discharge opening of the drum is raised and the mixture is discharged by tilting the drum.

tobermorite gel. The binder of concrete cured moist or in atmospheric pressure steam, a lime-rich gel-like solid containing lime in the ratio of 1.5 to 2.0 molecular weights of CaO to 1.0 molecular weight of SiO_2 (silica).

transit mix. Concrete in which the materials are charged into a truck mixer and mixing is done as the concrete is transported to the job site.

transit mixer. A truck that is mounted with a drum capable of mixing concrete.

traprock. Any of several kinds of dark-colored igneous rock.

trial batch. A batch of concrete prepared to establish or check proportions of the constituents.

trolley batcher. A batching plant consisting of two, three, or four hoppers for aggregates under which is suspended a movable weight hopper. This hopper can be positioned under each of the aggregate bins and quantities of each material can be accurately weighed.

The weigh bin can be arranged to charge directly into a mixer.

vermiculite. A group for certain platy minerals, characterized by marked exfoliation on heating. Sometimes used for aggregate in concrete.

vinsol resin. A natural or synthetic solid or semisolid organic material, used as an ingredient of air-entraining agents.

volumetric batching. The measuring of the constituent materials for mortar or concrete by volume.

warehouse pack. The stiffening of sacked cement stored for periods of time.

water reducer (water-reducing agent). A material that either increases workability of freshly mixed mortar or concrete without increasing water content or maintains workability with a reduced amount of water.

weigh batching. Measurement of quantities of materials by weight.

weighbeam. A beam on a scale containing movable weights that balance the materials in the weigh hopper. Size and positioning of these weights on the beam indicate the weight of material in the hopper.

weigh hoppers. A hopper or bin located beneath the cement and aggregate storage bins in a concrete plant. The weigh hopper is suspended from a scale system and is equipped with a discharge gate above the mixer.

wetting agent. A substance capable of lowering the surface tension of liquids, facilitating the wetting of solid surfaces and permitting the penetration of liquids into the capillaries.

workability. That property of freshly mixed concrete or mortar which determines the ease with which it can be mixed, transported, placed, compacted, and finished, without the loss of homogeneity.

APPENDIX

CONVERSION FACTORS—U.S. AND CANADIAN CUSTOMARY TO INTERNATIONAL SYSTEM OF MEASUREMENT (METRIC)*

To Convert From	To	Multiply By
Length		
foot	meter (m)	0.3048E†
inch	centimeter (cm)	2.54E
yard	meter (m)	0.9144E
mile (statute)	kilometer (km)	1.609
Area		
square foot	square meter (m^2)	0.0929
square inch	square centimeter(cm^2)	6.451
square yard	square meter (m^2)	0.8361

* This selected list gives practical conversion factors of units found in concrete technology. The reference source for information in Internal System of Measurement units and more exact conversion factors is "Metric Practice Guide" ASTM E380.
Symbols of metric units are given in parentheses.
† E indicates that the factor given is exact.

Volume (capacity)

cubic foot	cubic meter (m³)	0.02832
gallon (U.S. liquid)‡	cubic meter (m³)§	0.003785
gallon (Can. liquid)‡	cubic meter (m³)§	0.004546
ounce (U.S. liquid)	cubic centimeter (cm³)	29.57

Force

kilogram-force	newton (N)	9.807
kip	kilogram-force (kfg)	453.6
kip	newton (N)	4448
pound-force	kilogram-force (kgf)	0.4536
pound-force	newton (N)	4.448

Pressure or Stress (force per area)

kilogram-force/square meter	newton/square meter (N/m²)	9.807
kip/square inch (ksi)	kilogram-force/square centimeter (kgf/cm²)	70.31
pound-force/square foot	kilogram-force/square meter (kgf/m²)	4.882
pound-force/square foot	newton/square meter (N/m²)	47.88
pound-force/square inch (psi)	kilogram-force/square centimeter (kgf/cm²)	0.07031
pound-force/square inch (psi)	newton/square meter (N/m²)	6895

Bending Moment or Torque

inch-pound-force	meter-kilogram-force (m-kgf)	0.01152
inch-pound-force	newton-meter (Nm)	0.1130
foot-pound-force	meter-kilogram-force (m-kgf)	0.1383
foot-pound-force	newton-meter (Nm)	1.356
meter-kilogram-force	newton-meter (Nm)	9.807

Mass

ounce-mass (avdp)	gram (g)	28.35
pound-mass (avdp)	kilogram (kg)	0.4536
ton (metric)	kilogram (kg)	1000E
ton (short, 2000 lbm)	kilogram (kg)	907.2

Mass per Volume

pound-mass/cubic foot	kilogram/cubic meter (kg/m³)	16.02
pound-mass/cubic yard	kilogram/cubic meter (kg/m³)	0.5933
pound-mass/gallon (U.S.)†	kilogram/cubic meter (kg/m³)	119.8
pound-mass/gallon (Can.)†	kilogram/cubic meter (kg/m³)	99.78

Temperature**

deg Celsius (C)	kelvin (K)	$t_K = (t_C + 273.15)$
deg Fahrenheit (F)	kelvin (K)	$t_K = (t_F + 459.67)/1.8$
deg Fahrenheit (F)	deg Celsius (C)	$T_C = (t_F - 32)/1.8$

‡ One U.S. gallon equals 0.8327 Canadian gallon.
§ One liter (cubic decimeter) equals 0.001 m³ or 1000 cm³.
** These factors convert one temperature reading to another and include the necessary scale corrections. To convert a difference in temperature from Fahrenheit degrees to Celsius or Kelvin degrees, divide by 1.8 only, that is, a change from 70 to 88°F represents a change of 18°F or 18/1.8 = 10°C. To convert °C to °F, $t_F = 1.8t_C + 32$.

CONVERSION FACTORS FOR CONCRETE MIX DESIGN

Cement Content (per cubic yard)			Cement Content		Water-Cement Ratio		Water-Cement Ratio	
barrels	bags	pounds	pounds	bags	gal/bag	weight ratio	weight ratio	gal/bag
1	4	376	350	3.72	4	0.36	0.35	3.94
	4.5	423	400	4.25	4.5	0.40	0.40	4.50
1.25	5	470	450	4.79	5	0.44	0.45	5.07
	5.5	517	500	5.32	5.5	0.49	0.50	5.63
1.5	6	564	550	5.86	6	0.53	0.55	6.20
	6.5	611	600	6.39	6.5	0.58	0.60	6.76
1.75	7	658	650	6.92	7	0.62	0.65	7.32
	7.5	705	700	7.45				
2	8	752	750	7.98				
	8.5	799	800	8.52				
2.25	9	846						

Specify portland cement content and water-cement ratio by weight.

1 ton = 5.32 bbl (376 lb/bbl)
     = 21.28 bags (94 lb/bag)

METRIC EQUIVALENTS

Cement Content		Cement Content	
lb/cu yd	kg/m³	kg/m³	lb/cu yd
376	223	200	337
423	251	225	379
470	279	250	421
517	307	275	464
564	335	300	506
611	362	325	548
658	390	350	590
705	418	375	632
752	446	400	674
799	474	425	716
846	502	450	758
		475	801
		500	843

For	Multiply By
Cement content	
lb/cu yd to kg/m³	0.5922
Water-cement ratio	
gal/bag-to-weight ratio	0.0888

INDEX

multiple, 273–275
single-material weight type, 272–273
volumetric, 271–272
Batches, trial, *see* Trial batches
Batching, high-production, 271–275
on-the-job, 267–271
in quantities, 267–293
volumetric, 267–268
weighing tolerances for, 275–276
Beneficiation, 110
Bicarbonates, effects of in mixing waters, 96
Blade wear in mixers, 287–288
Bleeding, 175–176
Boats, 70
Bonding strength, tests of, 209
Bulking of aggregates, determination of, 114,
116

Carbonates, effects of in mixing waters, 96
Cast and vibrated pipe, 55
Cement, effect of on, strength, 198–200
workability, 177–178
history of, 11–23
minimum content of in absolute volume
method of proportioning, 247
portland, *see* Portland cements
special types of, 82–85
Cement batchers, 273
Central-mixing, 30
plants for, 290–292
Centrifugated pipe, 55
Channel improvements, 61
Charging, effect of on quality, 276–278
Chlorides, effect of on durability, 188
Cleanliness of aggregates, determination of,
124–125
Coarse aggregates, types of, 105
Coignet, Francois, 34
Compaction, 173–174
Compressive strength, tests of, 207
Concrete, air content of, 137–151
composition of, relation to proportioning,
220–223
durability of, 185–196
materials for, 73–161
mixing of, 212–293
properties of, relation to proportioning,
220–223
proportioning of, 212–293
quality of, 163–212

effect of admixtures on, 154
general properties of, 165
responsibility for, 166
strength of, 197–212
workability of, 173–183
Concrete industry, history of, 25–36
Concrete masonry, architectural uses of, 48–
49
Concrete-masonry industry, 26–28
Concrete pipe, water control uses of, 54–56
Concrete-pipe industry, 24–26
Concrete products, uses of, 37–71
Concrete storm cellars, 70
Concrete technology, 1–71
Concrete work, job opportunities in, 3–10
Conversion factors, 303–306
Cumulative batchers, 273–275

Dams, 59–61
De Navarro, Jose F., 22
Discharging of mixers, 285–286
Drainage, use of concrete products in, 68–69
Durability, of aggregates, 99–102
of concrete, 185–196

Economy, relation to aggregates, 104
Edison, Thomas A., 22
Elasticity, modulus of, tests of, 209
Entrainment of air, *see* Air entrainment
Expansive cements, 84

Fatigue strength, tests of, 210
Fineness of cement, 86
modulus of for aggregates in proportioning,
217–218
Fines, 171
Flexural strength, tests of, 208–209
Flow test for workability, 180–181
Formulas for aggregate properties, 113–125
Fountains, 70
Freeze-thaw conditions, effect of on durability,
193–194
Freezing, effect of on durability, of fresh
concrete, 193
of hardened concrete, 193–194
Fresh concrete, freezing of, 193
Freyssinet, Eugene, 35
Frost, James, 15
Full-pressure irrigation systems,
66

Open irrigation systems, 65–66
Oven-dry aggregate weight, determination of, 117
Overmixing, 281–283

Packerhead pipe, 55
Paint, Portland cement, 70
Parker, Joseph, 15
Particle-size distribution of aggregates, determination of, 113–114
Paving, features of, 38–43
Paving mixers, 289
Paving plants, on-site central-mixing, 292
Permanent mixers, 289–293
Permeability of liquids, effect of on durability, 186–187
Piling, 71
Plastic cements, 85
Poisson's ratio, determination of, 209
Pore moisture of aggregates, determination of, 119
Portable mixers, 288–289
Portland blast-furnace slag cement, 83
Portland cements, 75–91
  composition of, 75–77
  history of, 11–21
  manufacture of, 77–79
  modern history of, 12–17
  premodern history of, 12
  production practices for, 21–26
  setting time for, 87–89
  specification requirements for, 86
  types of, 79–82
  U. S. history of, 17–20
Portland-pozzolan cements, 83
Posttensioning, 35
Precast-concrete industry, 31–33
Precast concrete pipe, types of, 55
Pressure pipe, types of, 54–55
Pressure tests of air content, 142–148
Prestressed concrete, architectural uses of, 45–46
Prestressed-concrete industry, 35–36
Pretensioning, 35
Processing of aggregates, 105, 107, 109–111
Production of concrete, effect of on strength, 203–206
Production practices for Portland cement, 21–26
Proportioning, absolute volume method

of, 243–266
  and composition of concrete, 220–223
  of concrete, 212–293
  and fineness modulus of aggregates, 217–218
  with maximum density of aggregates, 216
  methods of, 223–231
  and properties of concrete, 220–223
  and surface area of aggregates, 216–217
  void content of coarse aggregate method of, 219–220
  voids-cement ratio, 218–219
Proportions, arbitrary, 216
  and optimum economic choice, 232–241

Quality of concrete, 163–212
  admixtures effect on, 154
  general properties of, 165
  responsibility for, 166
Quantity batching, 267–293
Quantity mixing, 267–293

Ransome, Ernest L., 34
Ransome, Frederick, 21, 34
Ready-mix plants, types of, 290–292
Ready-mixed concrete industry, 28–31
Recreation, uses of concrete products in, 69–70
Reinforced-concrete industry, 33–35
Remolding test for workability, 182–183
Retempering, 283–285
Rudolph, Paul, 49
Rural water supply, 57

Saturated surface-dry aggregate weight, determination of, 117
Saylor, David O., 18
Scales for wheelbarrows, 268–270
Sea-water, effect of on durability, 189, 192
  in mixing, 95–97
Segregation, 174–175
Self-stressing cements, 84
"Semiclosed" pipe irrigation systems, 66
Septic systems, 58
Setting time for Portland cements, 87–89
Sewage, effect of on durability, 189
  effect of in mixing waters, 96
Sewers, 57–58
Shell roofs, properties of, 43
Shipping of Portland cement, 91
Shore protection, uses of concrete

products in, 63
Shotcrete, 67
Shrink-mixing, 30, 291
Shrinkage, relation to aggregates, 103
Shrinkage-compensating cements, 84
Silt, effect of in mixing waters, 97
Single-material weigh batchers, 272–273
Skid resistance, relation to aggregates, 104
Slag cements, 83
Slump and overmixing, 282
Slump tests, in absolute volume method of
    proportioning, 246
  for workability, 178–180
Smeaton, John, 12
Sodium chloride, effects of in mixing waters,
    96
Soundness, of aggregates, determination of,
    122–123
  of Portland cements, 89
Special cements, types of, 82–85
Specific gravity of aggregates, determination
    of, 114
Specifications for Portland cements, 86
Spillways, 61
Stationary mixers, 289–290
Strain, 210
Straub, Francis J., 27
Strength, of concrete, 197–212
    effect on of materials, 198–203
    tests for, 207–211
  of Portland cements, 90
  relation to aggregates, 102–103
  relationships for, 211–212
Stress, 210
Storage, of aggregates, 111–112
  of Portland cement, 91
Sugar, effect of in mixing waters, 97
Sulfates, effect of on durability, 188
  effects of in mixing waters, 96
Surface moisture of aggregates, determination
    of, 119
Sutherland, C. C., 35

Tamped pipe, 55
Technicians, role of, 4–5
Technological training, need for, 5–10
Technology, growth of, 3–4
Temperature, effect of on workability, 178
Tensile strength, tests of, 207–208
Tests, nondestructive, 210–211

for strength of concrete, 207–211
of water for mixing, 94
of workability, 178–183
Thermal properties, relation to aggregates, 103
Time, effect of on workability, 178
Time of set for Portland cements, 87–89
Towers, 71
Trial batches, job-size, 240–241
  for unit-weight method of proportioning, 226
Trial mixes, absolute volume method of, 243–
    266
Trolley batcher plants, 270–271
Truck mixing, 30
Truck-mixing plants, 290–292
Tubes and risers, 69

Unit-weight method of proportioning, 225–232
  example of, 226–232
Unit weight of aggregates, determination of,
    114
United States, cement industry in, 19–23

Void content of coarse aggregate method in
    proportioning, 219–220
Voids-cement ratio method in proportioning,
    218–219
Volumetric air meters, 148–150
Volumetric batchers, 271–272
Volumetric batching, 267–268

Ward, William, 34
Waste water treatment, concrete products used
    for, 57–59
Water, effect of on strength, 200
  effect of on workability, 176
  impurities in, effects of, 93–94
  for mixing, impurities in, 94–97
  requirements for in absolute volume method
    of proportioning, 246–247
  tests of for mixing, 94
  uses of in concrete, 93–97
Water-cement ratios, 168–169
  in absolute volume method of proportioning,
    247–248
  effect of on strength, 206–207
Water control, concrete products used in, 53–
    59
Water restraint, use of concrete products in,
    59–65
Water supply, municipal, 56–57

rural, 57
Waterproofed Portland cement, 85
Weathering, effect of on durability, 186
Weighing tolerances in batching, 275–276
Wet aggregate weight, determination of, 118
Wettstein, Karl Bernhard, 35
Wheelbarrows, scales for, 268–270
White, Canvass, 17

White, John Bazley, 16
White Portland cement, 85
Wilkinson, W. B., 34
Workability of concrete, 173–183
  factors affecting, 176–178
  tests of, 178–183

Zeiss, Carl, 43